Did God Create the Universe from Nothing?

Countering William Lane Craig's Kalam Cosmological Argument

Jonathan M.S. Pearce

Foreword by

Jeffrey Jay Lowder

And with contributions from

James East & Counter Apologist

Did God Create the Universe from Nothing? Countering William Lane Craig's Kalam Cosmological Argument

Copyright © 2016 Jonathan M.S. Pearce

Published by *Onus Books*

Printed by Lightning Source International

Cover design: Onus Books and Jules Bailey. Image credits: Front cover - NASA, ESA, S. Beckwith (STScI), and The Hubble Heritage Team STScI/AURA); Back cover - NASA, ESA and the Hubble Heritage Team (STScI/AURA)

Trade paperback ISBN: 978-0-9926000-9-9

OB 12/18

PRAISE FOR *DID GOD CREATE THE UNIVERSE FROM NOTHING?*:

This is a beautifully crisp and clear introduction to, and discussion of, the Cosmological Argument. Suitable for beginners but also those who want a more insightful and detailed discussion. This is an ideal book for students, and indeed anyone who is interested in what remains one of the most popular arguments for the existence of God.

Stephen Law, Reader in Philosophy at Heythrop College, University of London and head of Centre for Inquiry UK.

The world is changing. Theism has lost its legs and has nothing upon which to stand, but the (failed) arguments in favor of the existence are still out there spurring and maintaining confusion. A needed remedy can be found in easily accessible, easily digestible rebuttals to the more popular apologetic arguments, and to this purpose, Pearce has again delivered, treating the important topic, the notorious (and bad) Kalam Cosmological Argument, in a concise and erudite way.

James A. Lindsay, PhD, Author of *Dot, Dot Dot: Infinity Plus God Equals Folly* and *Everybody Is Wrong About God*

Jonathan MS Pearce has done something remarkable. He has written an accessible, yet philosophically sophisticated, critique of the Kalam Cosmological Argument. He demonstrates familiarity with the latest philosophical literature on the Kalam, the origin of the universe, the nature of free will, and causality and time. What's more, he makes some novel contributions to this literature in the course of his analysis. If you have teethed yourself on popular discussions of atheism and religion, and now want to feast on something a little bit meatier, this is the book for you.

John Danaher, PhD, Lecturer in Law, NUI Galway (Ireland). Author of the blog *Philosophical Disquisitions*.

The Kalam argument enjoys much respect that it doesn't deserve, and *Did God Create the Universe from Nothing?* gives the unsparing rebuttal that it does deserve. Pearce is a capable and confident Virgil, guiding us through the philosophical and scientific fine points of the response. If you've read enough about Kalam to be intrigued and want the thorough takedown, this book is for you.

Bob Seidensticker, author of *Cross Examined: An Unconventional Spiritual Journey* and the "Cross Examined" blog at Patheos.com

With his latest book *Did God Create the Universe from Nothing?*, Jonathan Pearce has collected a vast array of the most powerful academic and popular-level responses to one of the most well-known cosmological arguments for the existence of God. Theists will be surely challenged by this wide-ranging book which seeks to put an end to this theistic argument about the beginning of the universe.

Justin Schieber, public debater on the philosophy of religion, creator of the channel *Real Atheology*

About the Author:

Jonathan MS Pearce is a philosopher, author, blogger, public speaker and teacher from Hampshire in the UK. His interests include, particularly, the philosophy of religion, which has remained the focus of his writing. Pearce's books include: *The Nativity: A Critical Examination*, *The Little Book of Unholy Questions* and *Free Will? An investigation into whether we have free will or whether I was always going to write this book*, as well as the ebook *The Problem with "God"* and editing and contributing to *13 Reasons To Doubt*. He lives with his twin boys and partner and wonders how she puts up with the three kids.

Acknowledgements:

This book was originally spawned some years back as the culmination of my philosophical studies at the University of Wales, Trinity St David, so I must thank them for providing the services they did. I am also very grateful to Counter Apologist, whom I cannot name properly because the U.S. still appears to hold atheists with a great deal of stigma! He has provided a valuable contribution, as has James East—many thanks to them both. Geoff Benson, as ever, has been a good source of second eyes. Always appreciated! The fantastic Jeffrey Jay Lowder has been a welcome addition to the project, and Bob Seidensticker's keen eyes and ideas have been truly valued. Jules Bailey was, for another time, a super help with the cover design. And again, Helen, my partner, deserves credit for her endless patience.

Thanks to all!

To Rob Stroud
...for helping to get
me into all of this

Contents

●

●

<u>Foreword</u>

There are many arguments for and against God's existence. For almost 40 years, philosopher William Lane Craig has promoted and defended one particular argument for God's existence more than any other: the kalām cosmological argument. In addition to his own body of work, there is now an entire body of secondary literature about the argument.

But why does the kalām cosmological argument continue to attract so much interest? There are probably many reasons. In no particular order, these reasons might include the following:

- Craig is widely regarded as one of the top, if not *the* top, defenders of theism today. He regularly debates the existence of God on college campuses around the world. When he does, his case for theism always includes the kalām cosmological argument, thus helping to provide the argument with broad exposure before the general public.

- As the old saying goes, "The best defense is a good offense." I suspect that many theists, including Craig himself, like the kalām cosmological argument because it enables theists to go on the offensive against people who say, "science disproves God." By appealing to scientific evidence for Big Bang cosmology in support of the kalām cosmological argument, the argument empowers theists to argue that scientific evidence supports the existence of God.

- Many people, myself included, find the argument interesting because it raises many important questions in philosophy (such as time, causation, realism, and free will), mathematics (e.g., transfinite arithmetic, infinity, etc.), science (i.e., cosmology), and theology (e.g., the relationship between God and time, whether the impossibility of an actual infinite provides the basis for an argument against Mormonism, whether a universe that is billions of years old is compatible with the age of creation implied by a literal interpretation of Genesis, etc.).

Did God Create the Universe from Nothing? is the best beginner- to intermediate-level book I've seen to critically assess the kalām cosmological argument. It covers a lot of the territory mentioned in my third bullet point above and so functions nicely as a sort of "user's guide" to the argument, but it does so in a way that should be accessible to the general reader.

In addition to its utility and accessibility, the book also appealed to me because of its intellectual integrity. Philosopher Quentin Smith once wrote an article in the journal *Philo* entitled, "The Metaphilosophy of Naturalism."[1] In that article, Smith argued that, while naturalism is true and theism is false, most contemporary atheist philosophers were *unjustified* in their belief that naturalism is true and theism false. Why? Because they accepted naturalism (and rejected theism) on the basis of bad arguments and objections. (It is sad to think about the fact that the naturalists who most need to read Smith's essay are the same naturalists who are also the least likely to do so.)

I view *Did God Create the Universe from Nothing?* as a helpful corrective to the trend described by Smith. Jonathan Pearce is to be commended for his obvious concern for accuracy; if only more atheists would read this book, then atheist commentators on the kalām argument would be much better informed. So I'll stop singing the book's praises now and let it go to press. Many readers interested in the argument—theists and nontheists alike—would greatly benefit from the insights of this book.

JEFFERY JAY LOWDER

PART ONE
The Background

1.1 The History

The Kalam[2] Cosmological Argument has a long and varied history. It is a member of a family of arguments known as cosmological arguments which set out to prove the existence of a First or Prime Cause (or Mover), usually entailing the existence of a personal god. The jump to a personal god is one which requires a lot more philosophy and work, and is something back to which I will refer later in this book. Of the Kalam Cosmological Argument (hereafter often referred to simply as the Kalam or KCA), Plato gave early reference to the idea:

> As for the world – call it that or cosmos or any other name acceptable to it – we must ask about it the question one is bound to ask to begin with about anything: whether it has always existed and has not beginning, or whether it has come into existence and started from some beginning. The answer is that it has come into being; for it is visible, tangible and corporeal, and therefore perceptible to the senses, and, as we saw, sensible things are objects of opinion and sensation.... And what comes into being or changes must do so, we said, owing to some cause. (Plato, *Timaeus*, 27-28)

The word *kalam*, in Arabic, means *discussion* or *discourse*, although its original inception seems to have been at the hands of Aristotle. Aristotle formulated an argument to show there to be an eternal universe. Early Christians attempted to refute this, and this was developed by Islamic theologians to argue for the existence of God.

3

The early thinking behind the argument ran along the lines that every motion needs a cause, and this regression of causality must have a beginning; hence the First Cause which does, itself, not require a cause. William Lane Craig, the philosopher on whom I will be concentrating in this book, quotes the Islamic scholar Al-Ghazali as saying[3]:

> "Every being which begins has a cause for its beginning; now the world is a being which begins; therefore, it possesses a cause for its beginning."

The argument has always been in the theistic toolbox, but it has not been until more recent times during a period of more intense debate between theists and non-theists in a more public context that the argument has been refined and utilised to the extent it now is. William Lane Craig, for example, is a modern proponent of the argument who uses the syllogism in almost every one of his public debates. It features prominently in his apologetic case for theism written for those wanting to argue for the existence of God: *A Reasonable Faith*. I will now introduce Craig.

1.2 William Lane Craig

William Lane Craig is one of the foremost proponents of the Kalam Cosmological Argument in recent times. Presently, he is Research Professor of Philosophy at Talbot School of Theology in La Mirada, California, having gathered philosophical and theological qualifications in the U.S., the U.K. and mainland Europe. Craig has authored or edited over thirty books, including *The Kalam Cosmological Argument* and the aforementioned *A Reasonable Faith* as well as contributing to many academic and religious journals. He has debated a tremendous number of people throughout his life, arguing most often for the existence of God, and reached some notoriety in Britain due to the fact that he challenged Richard Dawkins to debate him. Dawkins publicly refused.[4] Perhaps Craig's formidable experience and reputation as debater and orator acted as a deterrent. As Garrett J. DeWeese states in *Doing Philosophy as a Christian*[5]:

> William Lane Craig's scholarship has focused on defending the historicity of the resurrection of Jesus and on the kalam cosmological argument. His research and writing on the latter has sealed his reputation as a significant thinker on the philosophy of time, but he devotes a considerable portion of his time and efforts to apologetics, skilfully debating both philosophical and theological opponents of orthodox Christianity.

J.P. Moreland, Distinguished Professor of Philosophy also at the Talbot School of Theology, represents a view much held in the discipline of Christian apologetics[6]:

> It is hard to overstate the impact that William Lane Craig has had for the cause of Christ. He is simply the finest Christian apologist of the last half century, and his academic work justifies ranking him among the top one percent of practicing philosophers in the Western world.

High praise indeed. Craig has recently taken to defending the KCA in light of growing criticism of the argument from professional and amateur philosophers alike to the point that he has given public talks and contributed an essay to *Come Let Us Reason: New Essays in Christian Apologetics* entitled "Objections So Bad That I Couldn't Have Made Them Up" (variously titled as a public talk). It is this last writing and set of public talks that this book is targeting as I feel that Craig has wrongfully adopted a dismissive approach to the objections

5

to the KCA, evading the serious objections whilst building up straw men to other objections. I will detail this later in the book. Now let us look at the argument itself.

PART TWO
The Argument

This brings us on to the argument itself. I will formulate the argument as found in William Lane Craig's *A Reasonable Faith* so as not to create a straw man when dealing with Craig's arguments later in this book. As Craig declares[7]:

> The kalām cosmological argument may be formulated as follows:
> 1) Whatever begins to exist has a cause.
> 2) The universe began to exist.
> 3) Therefore, the universe has a cause.
> Conceptual analysis of what it means to be a cause of the universe then aims to establish some of the theologically significant properties of this being.

The further inference as to the properties of the initial cause (God) forms an extension to the argument, which Craig elucidates further. For the purpose of this book, this is of no concern here, other than a short note towards the end. I will be dealing exclusively with the two premises and the conclusion to this deductive argument. Thus, what is important is the concise version of the argument as listed above. Three simple steps leading to a First Mover as being responsible for the original causality of the universe. Let us look at the form of the argument. This will represent the first of my objections which I will set out in order of how they appear in the argument. After setting out the criticisms, I will then deal with objections to the arguments.

2.1 The Form

1) Whatever begins to exist has a cause.
2) The universe began to exist.
3) Therefore, the universe has a cause.

The above argument is a logical syllogism. A syllogism is an argument where the conclusion is inferred from the premises. As Craig likes to set out[8], it follows the same form as:

1) All men are mortal;
2) Socrates is a man;
3) Therefore, Socrates is mortal.

These two arguments are examples of *modus ponens* deductive arguments. Modus ponens syllogisms take on the symbolism:

$$P \rightarrow Q$$
$$P$$
$$\therefore Q$$

Which means that P implies Q; P is asserted to be true, and it follows, using the rule of inference, that Q is true. P and Q are categorical propositions. Both premises in both arguments contain a subject and a predicate. For example, in the Socrates syllogism, 2) contains the subject (Socrates) and the predicate (is a man) in the same way The Kalam Cosmological Argument second premise contains a subject (the universe) and a predicate (began to exist). These premises are essentially assertions which may or may not be true. The conclusion follows necessarily from these categorical propositions (premises). Thus the argument may be valid such that the conclusion *does* follow necessarily from the premises, but that the premises, as assertions, may be called into question.

This form of syllogism (modus ponens) is, as mentioned, a type of *deductive* reasoning. This means that, again as mentioned, the conclusion follows *necessarily* from the premises. As long as one agrees with the premises, then one cannot get away from the fact that the conclusion must be true. If, indeed, all men are mortal, and if, indeed, Socrates is a man, it follows without fail that Socrates is mortal. This is different to *inductive* logic, which entails garnering a conclusion based on premises that are built on patterns of individual instances, broadly speaking. In fact, it is a probabilistic method. Here is an example of inductive reasoning (as opposed to deductive reasoning):

1) All dogs that have ever been seen have ears;
2) Therefore, all dogs have ears.

The conclusion follows from the premise *probabilistically* and not *necessarily*. Describing the difference here is very important for looking at my first critique of the KCA. In the Socrates example, and as with all strong deductive arguments, the power of the logic is definitional. This is the key to deductive reasoning—the very definition of words determines whether or not the argument can be held to be true. As Nancy Cavender, Professor at the College of Marin, says of tautologies[9]:

> We can determine the truth of... a tautology by logical—deductive—means alone, without the need for any empirical investigation or inductive reasoning, But determining the truth or falsity of a contingent statement requires observation by ourselves and others and, usually, the employment of inductive reasoning.

Now, deductive arguments are not tautologies, not least because they are not coextensive such that the statements are not identical per se, but they *do* have an equivalence, if not in both directions. Built into the definitions of the words "man" and "mortal" is an equivalence (all men are mortals, even if all mortals are not necessarily men). A tautology would be like saying 3 = 3, whereas (though this analogy has its obvious limitations) a deductive argument might say 3 = 2 + 1. We accept the semantic meaning of both "man" and "mortal" so that, as Cavendar says, "without the need for any empirical investigation or inductive reasoning" we can deduce the conclusion. This is the strongest, most ironclad form of logic, resting on semantics as opposed to empirical observation and probability.

Let us, then, look at the KCA again. The first premise is "whatever begins to exist has a cause". This, I posit, is a categorical proposition which amounts to an inductive piece of reasoning akin to "all dogs that have ever been seen have ears" as opposed to a categorical proposition based upon definition such that "all men are mortal". The definition of "man" infers mortality. However, "whatever begins to exist has a cause" is not definitional and is dependent on empirical observation. This premise actually amounts to, at best, "everything which we have observed to begin to exist has appeared to have had a cause". It is an assertion, a mere assertion at that, that *everything* which begins to exist has a cause, in the same manner that the premise "all dogs that have ever been seen have ears". In the same way that it doesn't necessarily follow that "therefore, all dogs have ears", it doesn't necessarily follow that "everything that begins to exist has a cause". This can be called a *simple induction by enumeration to a generalisation.* Generalising a rule from a number of observations. Now, Craig might counter that this is not merely based on observation, but is also a metaphysical intuition. It just *feels* that this is true, that every effect has a cause. More on that later.

It is also worth bringing up here how deductive arguments can mask uncertainty. This is summed up by Jeffery Jay Lowder in a piece "How the Distinction between Deductive vs. Inductive Arguments Can Mask Uncertainty"[10]. Let's take a simple syllogism:

1) If A, then B
2) A
3) Therefore, B

What is the probability of B? Well, it is conditional upon A, so if A is true, then B has a probability of 1. It is certain. However, the *unconditional* probability of B is dependent on the probability of A. A deductive argument does a good job of *seeming* concrete, valid and sound, but it masks the weakness of the opening premise. Lowder states this in logical terms:

> But we want to know the unconditional probability of B, Pr(B), not the contribution made to Pr(B) by the probability of B conditional upon A, Pr(B|A). So, again, what is the contribution made to Pr(B) by A itself? Answer: Pr(A). Suppose A is true by definition. In that case, its probability is 1 and so Pr(B)=1. Now suppose the probability of A is 50%. In that case, Pr(B)=0.5. The probability calculus implies, when Pr(B|A)=1, that the contribution to Pr(B) made by A alone will always equal Pr(A), so long as Pr(A) is not zero, which in "real world" problems is usually the case.

Before I get on to talking at some depth about the idea of induction, let me say a little about acausality. There is a position that could be held, similar to Jung's synchronicity[11], that events might appear to be causally connected, appearing one before the other, but not *actually* having a causal relationship. This would be damaging to the position of adherents to the KCA. Strictly proving causality is notoriously difficult, and the theory of synchronicity is often seen as not being particularly useful.[12] Whilst humans cognitively over-determine causality in some situations, mistaking causality is not equivalent to claiming there is no causal relationship at all. For example, if I am in the jungle and see the undergrowth move, I may conclude that this is a tiger, and run away. Let's say I am wrong; it is not a tiger. This misattribution does not negate causality. It turns out, let us say, that it was the wind. Causality was still at play, just in a different form. As an aside, we can see why we may have evolved over-determining causality as it can work in our favour. Mistaking a wind for a tiger, and running away, I still live. That said, the idea that there *must* be causality is indeed arguable, and that certain events could be acausal yet synchronous is worth considering.

What my main objection to the form of the argument represents is the notion that within the deductive form of the KCA, the author smuggles in a premise which amounts to an *inductively garnered conclusion*. In other words, the KCA should be set out as follows:

1) Everything that we have observed to have begun seems to have had a cause
2) Therefore, everything that begins to exist has a cause
3) The universe began to exist
4) Therefore, the universe has a cause

However, this argument is *only* as good as its inductive premises (see my earlier point and Lowder's quote with regard to deductive arguments, too), and does not have the deductive qualities that many proponents proclaim. Of the form of the Socrates modus ponens, Craig claims, "This is one of the most basic and important logically valid argument forms"[13]. But his argument rests upon the idea that the KCA and the Socrates modus ponens arguments are synonymous in form when in fact they are ever so slightly, yet crucially, different.

This has a twofold implication. Firstly, as stated, its strength is only really inductive, allowing weakness into that first premise. Furthermore, the second premise is *also* not a premise dependent upon semantic logic, but upon empirically based observations and theorising. Secondly, Craig is using induction[14] to make a claim about things beginning to exist requiring a cause but he is inducing that information from things *in* the world, rather than inducing the premise from examples of worlds/universes coming into existence. Of that he has no other experience, and so even inductive knowledge can get Craig nowhere here. As philosopher Wes Morriston states in "A Critique of the Kalam Cosmological Argument"[15]:

> Is premise 1 of the *kalam* argument true? Must everything that begins to exist—even the very first event in the history of time—have a cause? Craig thinks it is unnecessary to give a lengthy defense of this claim. "Does anyone in his right mind", he asks, "really believe that, say, a raging tiger could suddenly come into existence uncaused, out of nothing, in this room right now?" [Craig 2002][16] Probably no one does. Craig then invites us to apply this "intuition" to the beginning of the universe, and the case for the first premise of the *kalam* argument is about as complete as Craig ever makes it.

> But surely this is much too quick? Of course, no one thinks a tiger could just spring into existence "in this room right now." But before we jump to conclusions, we need to ask *why* this is so. What makes this so obvious? Is it, as Craig seems to suppose, that all normal persons believe the first premise of the *kalam* argument, and then apply it to the case of the tiger? Call that the *top-down* explanation. Or is it rather that we have a lot of experience of animals (and other middle-sized material objects), and we know that popping up like that is just not the way such things come into existence? Call that the *bottom-up* explanation.

The bottom-up explanation takes note of the fact that we are dealing with a familiar *context* – one provided by our collective experience of the world in which we live and of the way it operates. It is our background knowledge of that context—our empirical knowledge of the natural order—that makes it so preposterous to suppose that a tiger might pop into existence uncaused. We *know* where tigers and such come from, and that just isn't the way it happens.

Now contrast the situation with regard to the beginning of time and the universe. There is no familiar law-governed context for it, precisely because there is nothing (read, "there is not anything") prior to such a beginning. We have no experience of the origin of worlds to tell us that *worlds* don't come into existence like that. We don't even have experience of the coming into being of anything remotely analogous to the "initial singularity" that figures in the big bang theory of the origin of the universe. The intuitive absurdity of tigers and the like popping into existence out of nowhere does not entitle us to draw quick and easy inferences about the beginning of the whole natural order.

Consequently, we have multiple weaknesses of the argument based on its logical form. I don't really want to go into detail here about the content of premise 1 within this context. Suffice it to say Craig is drawing inductive probability upon a behaviour, in premise 1, over which he has no probabilistic knowledge. Craig simply has no knowledge through experience of worlds beginning to exist (I will expand on this in section 3.4). As such, he has no epistemic right to claim the categorical proposition of premise 1 as being sound. Craig, in this case, merely appeals to intuition (but without the *relevant* evidence, since we have seen him use the ontology, or the nature of being, of one thing to assume the ontology of another) in a rather dismissive manner:

> First and foremost, the causal premiss is rooted in the metaphysical intuition that something cannot come into being from nothing. To suggest that things could just pop into being uncaused out of nothing is to quit doing serious metaphysics and to resort to magic. (Craig, 2007)

Thus we have an inductive premise which is based not on probabilistic empirical evidence (seeing many universes begin to exist with a cause) but on *intuition*. At the very least, this admits a weakness in the claims of deductive power imbued in the argument. Craig is relying on mere human intuition, and this argument is at best, then, an intuitive argument (albeit one that he uses observation to support). That does not necessarily invalidate it, but takes it down a step or two from the logical pedestal upon which Craig has placed it.

Did God Create a Universe from Nothing?

PART THREE
Premise 1

Having analysed the logical form of the argument, let us now move on to the first premise before going on to analyse the second premise. There are several objections to the first premise, but I will start off by looking at the problems with causality which critically affect the argument.

3.1 Causality making it a circular argument

What we need to think about first here is causality. Indeed, this whole argument is one over causality: cause and effect. Whilst cause and effect might be at face value a very simple thing, just the term "cause" can be tricky. When Craig talks about cause, he terms a cause as an *efficient cause* (Craig 1979) which is often defined as follows[17]:

> We can get some clarity on the question by recalling Aristotle's distinction between an efficient cause and a material cause. An efficient cause is something that produces its effect in being; a material cause is the stuff out of which something is made. Michelangelo is the efficient cause the statue David, while the chunk of marble is its material cause.

> If something popped into being out of nothing, it would lack any causal conditions whatsoever, efficient or material. If God creates something ex nihilo, then it lacks only a material cause. This is, admittedly, hard to conceive, but if coming into being without a material cause is absurd, then coming into being with neither a

17

material cause nor an efficient cause is, as I say, doubly absurd, that is, twice as hard to conceive. So it's not open to the non-theist confronted with the beginning of the universe to say that while creatio ex nihilo is impossible a spontaneous origin ex nihilo is.

The philosophy of causality is very complex, and indeed very dry, to the uninitiated. There are many different theories, of which Craig and Aristotle's is merely one. What, exactly, is being connected and how? Are they events, facts, features, states of affairs, situations, aspects, or so on?

I do not want to spend a whole book investigating the nature of causality, despite the fact that this is the most crucial concept to this argument, because it will derail the book, the flow and be counterproductive to the reader. Instead, I will adopt this approach. Causality means *something*. It conveys *some sense* that one circumstance in time is connected to another, and without the previous moment, or circumstance, taking place the latter one would not take place (in such a way). There is a vital connection and relationship between both. What this is and how it works is, therefore, not entirely necessary to set out. Rain makes crops grow. That we do not know what rain is (for example), and how it works, is not relevant for using the fact that there is *something* about rain that we can harness to make crops grow. I will leave it here, in its most simple form, in claiming that causality is something, it's not that it doesn't exist in some sense. Whatever that may be, is what we, both Craig and I, are referring to, without needing to explore the brickwork of the metaphysics. However, I will express a sense in which we fundamentally differ in the coming sections and how this is crucial in illustrating problems with his idea of the KCA.

With this in mind, let us look at causality and the problems with it. Let me analogise to make the point as clear as possible:

Smith is driving along the road over the speed limit. He is tired due to a heavy work schedule and a deadline which meant a lack of sleep the night before and is late for a meeting. One of his favourite songs comes on the radio and he starts singing along to it. On the pavement (sidewalk) a drunk man falls over into a bin which the Borough Council had just put in place to improve the cleanliness of the town. The bin is knocked off its stand and rolls into the road. Smith sees the bin late as his attention is distracted. He swerves, to avoid it. At the same time, a boy is trying to cross the road without looking. Smith is swerving into him and has to reverse his swerve significantly the other way, hitting a pothole in the poorly maintained road. This sends the car out of his control and onto the pavement. Jones, who had been walking by, slips on some soapy water draining from the carwash he is walking past. Whilst Jones is picking himself up, Smith's car mounts the pavement, hits Jones, and kills him instantly. What is the cause of Jones' death?

This is a very difficult, but standard causal question. The universe is not an isolation of one cause and one effect. It is a matrix of cause and effect with each effect being causal further down in something like the continuum. One could say that the impact of the car on Jones' head kills him. But even then, at what nanosecond of impact, what degree of the force killed him? This is arbitrarily cutting off the causal continuum at 1, half or quarter of a second before the effect (Jones' death). Having said that, the cause could be said to be the lack of oxygen to the brain, or the destruction of his vital organs. We could also accuse the bin, the drunk or anything else as being a cause, because without each of these, the final effect would not have taken place.[18]

As a result, I would posit that the cause of Jones' death is one long continuum which cannot be arbitrarily sliced up temporally.[19] As such, it stretches back to, say, the Big Bang—the start of the causal chain. In terms of free will, we call this the causal circumstance. Because the universe is one big causal soup, I would claim that any effect would be the makeup of the universe at any one point, like a snapshot. This makeup that leads to any given effect cannot be sliced up arbitrarily but is the entire connected matrix of 'causes and effect' (for want of a better term) since the Big Bang.

In other words, there is only one cause. The universe at the Big Bang (or similar).

If I am picking up a cup of tea to drink from it now, then we could just look at a few seconds before this as to the cause. Perhaps it was just my intention. But how about the notion that my parents introduced me to tea, and all those instances of tea drinking which came from that that now enforce my intentions? What if tea had not evolved? What if my grandparents had not given birth to my parents, and them to me? What if humanity had not evolved? What if the Earth had not harboured life? Without all of these, I would not have picked up my cup of tea. They are all relevant (and all the bits in between, and connecting them to other parts of the matrix) to my drinking tea now.

Philosopher Daniel Dennett uses another example about the French Foreign Legion that he himself adapted from elsewhere to illustrate problems with a basic notion of causality, of A simplistically causing B:

> Not that deadlocks must always be breakable. We ought to look with equanimity on the prospect that sometimes circumstances will fail to pinpoint a single "real cause" of an event, no matter how hard we seek. A case in point is the classic law school riddle:
>
> · Everybody in the French Foreign Legion outpost hates Fred, and wants him dead. During the night before Fred's trek across the desert, Tom poisons the water in his canteen. Then, Dick, not knowing of Tom's intervention, pours out the (poisoned) water and replaces it with sand. Finally, Harry comes along and

pokes holes in the canteen, so that the "water" will slowly run out. Later, Fred awakens and sets out on his trek, provisioned with his canteen. Too late he finds his canteen is nearly empty, but besides, what remains is sand, not water, not even poisoned water. Fred dies of thirst. Who caused his death?

This thought experiment defends the thesis that causality is, at times, impossible to untangle or define. I would take this one very large step further in saying that the causality of such an effect, of *any* effect, is traceable back to the first cause itself: the Big Bang or whatever creation event you ascribe to.[20]

So the causality of things happening now is that initial singularity or creation event. As I will show later, nothing has begun to exist, and no cause has begun to exist, other than that first cause—the Big Bang singularity.

Let me show this as follows with another example of such causality. In this example, the term causal circumstance is the causal situation that has causal effect on the object—from every air molecule to every aspect of physical force:

Imagine there are 5 billiard balls A–E and nothing else. These came to exist at point t_0 with an 'introductory force'. At each point t_1, t_2 etc, every ball hits another ball. At point t_5, B hits E at 35 degrees sending it towards C. Craig's own point about causality seems to be this: the cause for B hitting E at 35 degrees is the momentum and energy generated in B as it hits E. That is his 'efficient cause'. My point is this: the cause of B hitting E is at t_0. No cause has begun to exist or has been created out of nothing. The causes transform— what is called transformative creation. So the cause of B hitting E is:

B firing off at t_0 and hitting A at t_1, the causal circumstance meaning it rebounds off A to hit D at t_2, meaning the causal circumstance rendering it inevitable that it hits A again at t_3... and then it hits E at t_5 at 35 degrees.

The cause is the casual circumstance at t_5. This is identical to the causal circumstance in free will discussions—that determinism entails the cause of an action to the first cause of the Big Bang. The causal circumstance is everything up until the moment t_5 as well as all the factors at the moment just prior to t_5 (at t_4). Craig is incorrect, in my opinion, in saying that the cause of B hitting E is the immediate isolated efficient cause just before t_5 (t_4).

Now, in this example, the term *transformative creation* pops up. This is something which will be examined in the next objection. In summary, this example shows that one cannot arbitrarily quantise causality; one cannot cut it up into discrete chunks since it is, in reality, one long, continual causal chain, unbroken. Even the word "chain" is problematic because it is linear in sense, and causality is more like a matrix (or causal soup as I earlier said!).

What this amounts to is the notion that there is only one cause, and even this is open for debate. The creation event sets in motion one long, interconnected continuum of causality. What I am implying here, then, is that there is only one effect. This means that the idea that "everything" or "every effect" as it can be synonymously denoted is incoherent since there is only one effect—an ever morphing matrix of causality. Let us see how this changes the syllogism:

1) Everything which begins to exist has *the universe as the causal condition for its existence.*

2) The universe began to exist.

3) *Therefore, the universe had the universe as a causal condition for its existence.*

As you can easily see, the conclusion is highly problematic. It is nonsensical and seems to insinuate that the universe is self-caused. There is only one cause and this is the universe and can hardly be applied to itself. One cannot make a generalised rule, which is what the inductively asserted first premise is as we have discussed, from a singular event/object and then apply the rule to that very event/object. This is entirely circular and even incoherent. Causality itself renders the KCA problematic.

In exactly the same way that we cannot untangle or slice up causality into discrete parts, we cannot also delineate objects, and this leads me on to the second objection which closely matches the one just elucidated.

3.2 Nominalism and "everything" being "the universe"

Authors of the KCA, such as Craig, see the argument as dealing with the beginning of existence of all discrete objects as being the set described by the term "everything". In other words, a chair, a marble, a dog and a mountain all begin to exist and have causes for their respective existences. This would be, admittedly, the commonsense understanding of the ontology of these objects—that they begin to exist at a particular point in time from having not existed at a previous point in time. What I am going to set out is very similar to one of Adolf Grünbaum's objections that he set out in his 1990 essay "The pseudo-problem of creation in physical cosmology ".[21]

The problem for the KCA is the definition of "everything". My claim is that everything is in fact "the universe" itself. As Grünbaum states:

> ...consider cases of causation which do involve the intervention of conscious fashioners or agents, such as the baking of a cake by a person. In such a case, the materials composing the cake owe their particular state of being in cake-form partly to acts of intervention by a conscious agent. But clearly, the very existence of the

atoms or molecules composing the cake cannot be attributed to the causal role played by the activity of the agent. Thus, even if we were to assume that agent-causation does differ interestingly from event-causation, we must recognize that ordinary agent-causation is still only a *transformation* of matter (energy)....

Even for those cases of causation which involve conscious agents or fashioners, the premise does not assert that they ever create anything *out of nothing*; instead, conscious fashioners merely TRANSFORM PREVIOUSLY EXISTING MATERIALS FROM ONE STATE TO ANOTHER; the baker creates a cake out of flour, milk, butter, etc., and the parents who produce an offspring do so from a sperm, an ovum, and from the food supplied by the mother's body, which in turn comes from the soil, solar energy, etc. Similarly, when a person dies, he or she ceases to exist *as a person*. But the dead body does not lapse into nothingness, since the materials of the body continue in other forms of matter or energy. In other words, all sorts of organization wholes (e.g., biological organisms) do cease to exist only *as such* when they disintegrate and their parts are scattered. But their parts continue in some form.

We can, here, start to see an issue with the idea, in the first premise, of things beginning to exist with the notion of transformative creation as mentioned previously. We have already discussed how all causes can be reduced to a single cause. Now I will set out, as Grünbaun hints at, to show that "everything" is a term which also refers to a singular object.

Firstly, the only thing, it can be argued, that "has begun to exist" is the universe itself (i.e. all the matter and energy that constitute the universe and everything in it). Thus the first premise and the conclusion are synonymous—the argument is entirely circular.

So how do I establish that the only thing which has begun to exist is the universe? We may think that things like tables, chairs, humans, rocks, lemmings and so on exist. Well, they do in one sense (an arrangement of matter/energy), but in the sense of the abstract labels of "rock" or "chair", they are exactly that, abstract labels. Their existence, in Platonic terms, as some kind of objective entity, requires the philosophical position of (Platonic) realism. Platonic realism, in simple terms, is the position that universals such as redness or doghood and abstractions (kinds, characteristics, relations, properties etc.) are not spatial, temporal or mental but have a different ontology, existing separately from the objects which instantiate such properties.[22] The opposite position to this is nominalism, which can mean the denial of the existence of these abstract labels in some sense.

For example, in order for the statement "John Smith is a gardener" to hold a truth value, there must be some existence property defined by "gardener" such as "gardenership". This universal is different from the instance of the universal property found in John Smith. This is not a position that Craig adheres to. All we have on a nominalist or conceptualist worldview (as opposed to realist) is a transformative coming into existence. What this means is that what makes the chair, the molecules and atoms, already existed in some form or other before the "chair" came to be. So the matter or energy did not "begin to exist". This merely leaves the label of "chair".

The nominalist, as stated, adopts a position which denies the existence of universals, such as redness or gardenership, and claims that only individuals or particulars exist. Conceptualism or conceptual nominalism, on the other hand, is a position which claims that universals only exist within the framework of the thinking (conceiving) mind. Most philosophers agree that abstract objects are causally inert, by definition. This means that, at best, the abstract label is unable to have causal power anyway (regardless of its ontology).

To illustrate this, let's now look at the "label" of "chair" (in a very cogent way, all words are abstractions that refer to something or another, but nominalists will say that these abstractions, or the relationship between them and the reference points, do not exist, out there, in the ether). This is an abstract concept, I posit, that exists, at most, only in the mind of the conceiver. We, as humans, label the chair abstractly and it only means a chair to those who see it as a chair—i.e. it is subjective. The concept is not itself fixed. My idea of a chair is different to yours, is different to a cat's and to an alien's, as well as different to the idea of this object to a human who has never seen or heard of a chair (early humans who had never seen a chair, for example, would not know it to be a chair. It would not exist as a chair, though the matter would exist in that arrangement). I may call a tree stump a chair, but you may not. If I was the last person (or sentient creature) on earth and died and left this chair, it would not be a chair, but an assembly of matter that meant nothing to anything or anyone.[23] The chair, as a label, is a subjective concept existing in each human's mind who sees it as a chair. A chair only has properties that make it a chair within the intellectual confines of humanity. These consensus-agreed properties are human-derived properties, even if there may be common properties between concrete items— i.e. chairness. The ascription of these properties to another idea is arguable and not objectively true in itself. Now let's take an animal— a cat. What is this "chair" to it? I imagine a visual sensation of "sleep thing". To an alien? It looks rather like a "shmagflan" because it has a "planthoingj" on its "fdanygshan". Labels are conceptual and depend on the conceiving mind, subjectively.

What I mean by this is that *I* may see that a "hero", for example, has properties X, Y and Z. *You* may think a hero has properties X, Y and B. Someone else may think a hero has properties A, B and X.

Who is right? No one is right. Those properties exist, in someone, but ascribing that to "heroness" is a subjective pastime with no *ontic* reality, no *objective* reality.

This is how dictionaries work. I could make up a word: "bashignogta". I could even give it a meaning: "the feeling you get when going through a dark tunnel with the tunnel lights flashing past your eyes". Does this abstract idea not objectively exist, now that I have made it up? Does it float into the ether? Or does it depend on my mind for its existence? I can pass it on from my mind to someone else's using words, and then it would be conceptually existent in two minds, but it still depends on our minds. What dictionaries do is to codify an agreement in what abstract ideas (words) mean, as agreed merely by consensus (the same applies to spelling conventions—indeed, convention is the perfect word to illustrate the point). But without all the minds existing in that consensus, the words and meanings would not exist. They do not have Platonic or ontic reality.

Thus the label of "chair" is a result of human evolution and conceptual subjectivity (even if more than one mind agrees).

If you argue that objective ideas do exist, then it is also the case that the range of all possible entities must also exist objectively, even if they don't exist materially. Without wanting to labour my previous point, a "forqwibllex" is a fork with a bent handle and a button on the end (that has never been created and I have "made-up"). This did not exist before now, either objectively or subjectively. Now it does—have I created it objectively? This is what happens whenever humans make up a label for anything to which they assign function etc. Also, things that other animals use that don't even have names, but to which they have assigned "mental labels", for want of better words, must also exist objectively under this logic. For example, the backrubby bit of bark on which a family of sloths scratch their backs on a particular tree exists materially. They have no language, so it has no label as such (it can be argued that abstracts are a function of language). Yet even though it only has properties to a sloth, and not to any other animal, objectivists should claim it must exist objectively. Furthermore, there are items that have multiple abstract properties which create more headaches for the objectivist. A chair, to me, might well be a territory marker to the school cat. Surely the same object cannot embody both objective existences: the table and the marker! Perhaps it can, but it just seems to get into more and more needless complexity.

When did this chair "begin to exist"? Was it when it had three legs being built, when 1/2, 2/3, 4/5, 9/10 of the last leg was constructed? You see, the energy and matter of the chair already existed. So the chair is merely a conceptual construct. More precisely a human one. More precisely still, one that different humans will variously disagree with.

Let's take the completed chair. When will it become not-a-chair? When I take 7 molecules away? 20? A million? This is sometimes called the paradox of the beard / dune / heap or similar. However, to be more correct, this is an example of the Sorites Paradox, attributed to Eubulides of Miletus. It goes as follows. Imagine a sand dune (heap) of a million grains of sand. Agreeing that a sand dune minus just one grain of sand is still a sand dune (hey, it looks the same, and with no discernible difference, I cannot call it a different category), then we can repeatedly apply this second premise until we have no grains, or even a negative number of grains and we would still have a sand dune. Such labels are arbitrarily and generally assigned so there is no precision with regards to exactly how many grains of sand a dune should have.

This problem is also exemplified in the *species problem* which, like many other problems involving time continua (defining legal adulthood etc.), accepts the idea that human categorisation and labelling is arbitrary and subjective. The species problem states that in a constant state of evolving change, there is, in objective reality, no such thing as a species since to derive a species one must arbitrarily cut off the chain of time at the beginning and the end of a "species'" evolution in a totally subjective manner. For example, a late Australopithecus fossilised skull could just as easily be labelled an early Homo skull. An Australopithecus couple don't suddenly give birth to a Homo species one day. These changes take millions of years and there isn't one single point of time where the change is exacted. There is a marvellous piece of text that you can see, a large paragraph[24] which starts off in the colour red and gradually turns blue down the paragraph leaving the reader with the question, "at which point does the writing turn blue?" Of course, there is arguably no definite and objectively definable answer—or at least any answer is by its nature arbitrary and subjective (depending, indeed, on how you define "blue").

So, after all that, what has begun to exist? A causally inert abstract concept.

You see, once we strip away the labels and concepts, all we have left is matter and energy which is only ever involved in what has been called transformative creation, meaning it doesn't begin to exist, but is being constantly reformed throughout time. It only *began* to exist at the Big Bang or similar (in Craig's model).

So where does this leave us? The implications are twofold. Firstly, as Grünbaum illustrates, with all effects being merely transformative creations (i.e. nothing comes into existence but is transformed from already existing matter or energy), then we have an equivocation of the term cause. In Premise 1 we are talking about transformative causality, whereas in the conclusion we are talking about *creation ex nihilo* or creation out of nothing. As Grünbaum reasons[25]:

> Since the concept of *cause* used in the conclusion of the argument involves creation *out of nothing*, we see that it is plainly different from the concept of cause in the premise. And for this reason alone, the conclusion does not follow from the premise deductively.

This amounts, then, to a *fallacy of equivocation* whereby the author is using two distinct meanings of the same term in a syllogism. This makes the argument logically invalid or fallacious.

The second ramification of this line of argument is that it means that the term "everything" is actually synonymous with "the universe", with the universe being a set of finite energy and matter that has remained, in accordance with the Law of the Conservation of Energy, constant over time. We have agreed, then, that abstract concepts might begin to exist, but these are causally inert and do not exist objectively—only in the minds of the conceiver. One can then take this a step further and claim that, for a whole host of reasons (most of which I will not get into now), mental conceptions supervene on physical matter. That means that my mental states, and all the abstract concepts which they obtain, depend on the physical. One simple way of knowing this is whether my mental concept of a chair remains the same if I was to stick a fork into my eye and through into my brain. Our consciousness, in some way, is dependent on our brains states and matter. If you don't believe me, try it out.

So that leaves matter and energy, which have existed for all of time because they are, in effect, the universe itself (as is time, when understood as spacetime). It is not that the universe is "made up" of lots of matter and energy making it something, it simply *is* a quantity of matter and energy. We can refer back to our previous talk of conceptual nominalism. The "universe" is not some distinct thing from what it is made up. "Universe" is an abstract concept made up by humans to refer to "everything". Everything in existence that we can observe, that we can infer, and so on. This has some fairly crucial implications for the KCA necessitating a reformulation as follows:

1) The universe that begins to exist has a cause for its existence;
2) The universe begins to exist;
3) Therefore, the universe has a cause for its existence.

If we then project the syllogistic changes from Section 3.1 over this reformulated syllogism then we get an even more tautologous and incoherent argument:

1) The universe that begins to exist has the universe as the causal condition for its existence.

2) The universe began to exist.

3) Therefore, the universe had the universe as a causal condition for its existence.

As we can plainly see, if we delve into the actual meaning of these terms and input these definitions back into the syllogism we are presented with an argument that amounts to little more than nonsense.

One could claim, however, that this argument relies at least partially on the establishment of nominalism, conceptualism or some other form of non-realism in order to work. To this we shall now turn.

3.3 Establishing a non-realist position

There is not the time or the space to dedicate to properly establishing any such ontology here since much ink has been spilled and books written on realism, nominalism and everywhere in between. I will endeavour, though, to sketch out some basic arguments for a non-realist position in order to show that such a position is at least coherent, and more likely rationally preferable.

Firstly, I have already done a lot of spadework in the above section, which to me shows some robust defence of the conceptualist position and renders the position of Platonic realism problematic.

Before continuing to establish this, though, it is important to bring up an objection to authors of the argument such as William

Lane Craig. Craig, for example, uses the KCA on a regular basis both in print and within the realms of public debate and argument. However, I have never seen or heard him deal specifically with this matter in those public forums. What this means is that, when presenting the argument, in particular to the philosophical layman who would be unable to notice such a situation, Craig implicitly assumes a realist position (believing in some kind of Platonic realm of abstract ideas existing outside of our minds), or at least does not set out a case against nominalism or conceptualism in this context. This means that there are unreferenced philosophical assertions that undergird the entire argument, which Craig simply ignores or neglects to communicate to his audiences. Even if such positions as nominalism *aren't* philosophically tenable, it is at least his duty in claiming such rational power in the argument to outline such a case in order to show that the argument sits on firm philosophical foundations. To ignore such reference is to arouse suspicion within genuine philosophical circles (and this was indeed a large motivation for starting this project). And if the case for such positions as nominalism *is* philosophically tenable, then the burden of proof is on Craig and other KCA authors to defend the KCA from such objections. Admittedly, any philosophical position is doubtlessly built upon further philosophical assumptions, theories and notions, but this area is of such crucial importance to the argument that it simply cannot be ignored.

It is necessary to illustrate a coherent framework for a non-realist position to show that the argument I have provided is not simply a castle in the air. In order to do this, I will provide outlines of some arguments against universals and abstracts and I will then present the basis for a positive case for a non-realist approach.

To set the case out, let me define some terms first. An abstract object is notoriously difficult to define since it is easy to say what it is not, yet very hard to say what one actually *is*. It is generally, though not universally (such is the way of philosophy!), accepted that abstracts are non-spatiotemporal and causally inert objects meaning

32

they cannot have any causal effect and have no existence within the framework of space and time. A universal can be defined as follows (from the Stanford Encyclopedia of Philosophy)[26]:

> Realists about universals typically think that properties (e.g. whiteness), relations (e.g. betweenness), and kinds (e.g. gold) are universals. Where do universals exist? Do they exist in the things that instantiate them? Or do they exist outside them? To maintain the second option is to maintain an *ante rem* realism about universals. If universals exist outside their instances then it is plausible to suppose that they exist outside space and time. If so, assuming their consequent causal inertness, universals are abstract objects. To maintain that universals exist in their instances is to maintain an *in re* realism about universals. If universals exist in their instances, and their instances exist in space or time, then it is plausible to think that universals exist in space or time, in which case they are concrete. In this case universals can be multiply located, i.e. they can occupy more than one place at the same time, for *in re* universals are wholly located at each place they occupy (thus if there is whiteness *in re*, then such a thing can be six meters apart from itself).
>
> Thus, both on *ante rem* and *in re* realism about universals, universals enjoy a relation with space very different from that apparently enjoyed by ordinary objects of experience like houses, horses and men. For such particulars are located in space and time and cannot be located in more than one place at the same time. But universals are either not located in space or else they can occupy more than one place at the same time.

33

A nominalist denies existence of one or both of these things (abstracts and universals) and conceptualists claim they exist only in the minds of conceivers. The first argument against them is Ockham's Razor—the simplest explanation (one without the need of unnecessarily multiplying entities) is often the correct one. In this case, if one was able to show that concrete (i.e. non-abstract) objects can do the job of abstract objects, then this explanation is more attractive than the explanation that posits both concrete and abstract objects.

A classic argument in this realm of the Problem of Universals is the location of universals. What loci do these universals inhabit? Is there some Platonic realm or dimension where these abstracta reside or are they instantiated in this dimension? Given their non-spatiotemporal property, this is problematic.

Moreover, if abstracts are causally inert, then how can we have knowledge and belief about them, how can we talk or think about them? Perhaps there is a feedback loop such that knowledge of the abstracts has causal power.

Other objections include an interesting criticism known as Bradley's Regress. Relations are abstract objects such that when an entity instantiates a universal, there is an instantiation relation between them. For example, there is a relation between whiteness and a white cup. However, since there are many white cups, then this relation becomes a relational universal. This means that when a cup instantiates whiteness, then that relation and the cup and whiteness are linked by another instantiation relation. This relational instantiation goes on and on ad infinitum as we find relationships between relationships, and then further relationships and so on, thus causing an infinite known as Bradley's Regress. As Craig would himself claim, actual infinite sets are impossible.

There are other arguments against universals which are intricate and demand more time and space than I will allow. As such, I will refer the reader to Rodriguez-Pereyra (2015) for further reference.

As far as establishing a positive case for nominalism, as opposed to simply illustrating the problems with universals and abstracts without explaining their ontology, I will present a few brief cases. Firstly, *trope nominalism* or *trope theory* states that the universals are in fact particulars which are instantiated in every individual object. In other words, the whiteness of the cup is actually not a universal whiteness but a particular whiteness of *that* cup. *That* whiteness only exists in *that* cup. The fact that other white cups seem to have that property is that they resemble each other.[27]

This theory is similar to something called *Resemblance Nominalism* which differs from trope theory and does not supposedly fall into similar problems[28]:

> A final version of Nominalism is Resemblance Nominalism. According to this theory, it is not that scarlet things resemble one another because they are scarlet, but what makes them scarlet is that they resemble one another. Thus what makes something scarlet is that it resembles the scarlet things. Similarly, what makes square things square is that they resemble one another, and so what makes something square is that it resembles the square things. Resemblance is fundamental and primitive and so either there are no properties or the properties of a thing depend on what things it resembles.
>
> Thus on one version of the theory a property like *being scarlet* is a certain class whose members satisfy certain definite resemblance conditions. On another version of the theory there are no properties, but what makes scarlet things scarlet is that they satisfy certain resemblance conditions.

35

And there are other theories too. Suffice it to say that there are good cases for abstracta and universals being explicable through nominalism, and that is not even presenting a case for conceptualism or even idealism in explaining (away) the Problem of Universals.

It gets a little more complex when looking at Craig's position, because he has ended up changing it over time:

> A good place to begin is by asking ourselves, "Does the number 3 exist?" Certainly there can be three apples, for example, on the table; but in addition to the apples does 3 itself exist? We're not asking whether the numeral "3" exists (the symbol borrowed from the Arabs to represent the quantity three). Rather we're asking whether the number 3 itself exists. Are there such things as numbers? Do numbers really exist?
>
> Some people might think that this question is so airy-fairy as to be utterly irrelevant. But in fact it raises a fundamental theological issue whose importance can scarcely be exaggerated. For if we say that numbers do exist, where did they come from? Christian theology requires us to say that everything that exists apart from God was created by God (John1:3). But numbers, if they exist, are almost always taken to be necessary beings. They thus would seem to exist independently of God. This is the view called Platonism, after the Greek philosopher Plato.
>
> Someone might try to avoid this problem by espousing a modified Platonism, according to which numbers were necessarily and eternally created by God. But then a problem of vicious circularity arises: explanatorily prior to God's creating the number 3, wasn't it the case that the number of persons in the Trinity was 3? Of

course; but then the number 3 existed prior to God's creating the number 3, which is impossible!

I remember the sense of panic that I felt in my breast when I first heard this objection raised at a philosophy conference in Milwaukee. It seemed to be an absolutely decisive refutation of theism. I didn't see any way out.

This seemingly dry and often-ignored area of philosophy has fundamental ramifications for theology and the philosophy of religion. The realism of number presents a real problem for God. As such, after much discussion in the article from which the quote above comes, Craig concludes[29]:

In sum, the abundance of Nominalist defeaters of the Indispensability Argument leaves the issue of the ontological status of abstract objects like numbers at least an open question and various Nominalisms (not to speak of Conceptualism) as viable alternatives to Platonism. Hence, I'm pleased to say, no successful objection to classical theism arises from this quarter.

In other words, on numbers at least, Craig favours a type of nominalism. Much more recently, it appears that Craig has further adapted his thought[30]:

Matthew, your question illustrates exactly the reason I have avoided the term "nominalism" as a label for my views on the reality of abstract objects and chosen "anti-realism" instead.

In the history of theology "nominalism" has become something of a dirty word because of its associations with conventionalism and relativism, just as you suggest. But that is no part of contemporary anti-Platonism. Just as saying, "Mars has two moons" does not commit you to the existence of some weird abstract object we call 2, so saying, "Marriage is essentially (or by definition) a union between a man and woman" does not commit you to some abstract object which we call the essence of marriage. "The essence of marriage" is not meant to be a heavyweight, metaphysical term, but convenient shorthand for talking about the way marriage necessarily is.

In fact, in his paper "Nominalism and Divine Aseity", William Lane Craig sets out a defence of theistic nominalism[31]:

This chapter argues that if a Christian theist is to be a Platonist, then, he must, it seems, embrace Absolute Creationism, the view that God has created all the abstract objects there are. Those of us who find the boot-strapping problem compelling, however, must look elsewhere to find some solution to the problem posed by the existence of uncreatables. In recent decades there has been a proliferation of nominalistic treatments of abstract objects which has served to make Nominalism an attractive alternative for the orthodox theist. Van Inwagen himself holds that there is rightly a strong presumption of Nominalism's truth which only a rationally compelling argument for Platonism can overcome. Even if we do not hold to such a presumption, the orthodox Christian who is not an Absolute Creationist has grounds for thinking that Platonism is false and therefore has powerful reasons for entertaining Nominalism. Unless all forms of Nominalism can be shown to be untenable, the orthodox Christian can on theological grounds

rationally embrace Nominalism as a viable alternative
to Platonism.

In conclusion to this section, I posit that Craig and other authors
of the KCA are required to deal with such metaphysical postulations
in order to be able to hold to the notion that things begin to exist or
are (discretely) causes or themselves be causes when these objects
only appear to be, at best, causally inert abstracta. The problems of
universals and particulars cause a problem from KCA. Some might
claim it does not matter whether we can categorise things, it only
matters that things exist, but if these problems conclude that the only
things that have causal existence are things of matter and energy, and
not abstracta, then this changes what the KCA indeed refers to.

To make matters worse for Craig is the fact that he appears to
now be embracing nominalism far more, without, as far as I can tell,
showing how this has affecting his favourite argument. In other
words, Craig might well agree with the claims that I have made above,
concerning nominalism and conceptualism. This should worry him.

3.4 The Kalam Cosmological Argument and Libertarian Free Will are incompatible

Essentially, the KCA implies by assertion that you cannot have
ex nihilo creation—that an event cannot be created out of nothing,
and by nothing. Only God, Craig and others suppose, can do this.
And thus the universe, created out of nothing (arguably) was created
ex nihilo by God. The causal chain goes back to the Big Bang and
stops. How can this be explained? Well, since causality must continue
regressing backwards, if there is a beginning to causality, it can *only*
be explained by God.

However, the theist is usually, if not a Calvinist, an adherent to the notion of libertarian free will (LFW). By this, I mean that they believe an agent could make free choices—could have done otherwise. This implies that the agent is the *originator* of a freely willed decision, or the causal chain in a decision. The determinist, on the other hand, believes that every effect has a cause and that *that* cause is itself an effect or a prior cause, and this goes back to the Big Bang or similar starting point where physics breaks down, or some other such situation (quantum cosmological loop or suchlike). As Roderick Chisholm claims[32] that when we act freely, we exercise:

> ...a prerogative which some would attribute only to God: each of us, when we act, is a prime mover unmoved. In doing what we do, we cause certain things to happen, and nothing—or no one—causes us to cause those events to happen.

You cannot get clearer language that ties libertarian free will to the Kalam in insisting on there being uncaused causation. Chisholm might not claim that God is the only prime mover, but most theists, including Craig, certainly do. This is the talk of a prime mover, and God is supposed to be the only one. In case you are doubting this, here is Robert Kane, the famous naturalist philosopher rare in his insistence on the existence of LFW[33]:

> Free will...is the power of agents to be the ultimate creators or originators and sustainers of their own ends or purposes...when we trace the causal or explanatory chains of action back to their sources in the purposes of free agents, these causal chains must come to an end or terminate in the willings (choices, decisions, or efforts) of the agents, which cause or bring about their purposes.

The point is that the denier of LFW claims that the agent is themselves part of a larger causal chain which explains why the agent did what they did; that the reasons were derived from what is known as the causal circumstance—the snapshot of the universe at that prior moment to the event.

On the other hand, the theist generally believes that the causal chain starts with the agent; that they *originate* the causal chain. This allows them ownership over the decision so that the reason for the decision cannot be further deferred to other (antecedent) causes. However, this means that the agent is creating something out of nothing. There is ex nihilo creation, since no prior reason can be given to explain the agent's decision, otherwise we return to determinism.

Yet allowing for ex nihilo creation defies the opening premise of the KCA. William Lane Craig is always espousing the intuitive "truth" (and we know how unreliable intuition can be) of the metaphysical claim that *ex nihilo nihil fit*—out of nothing, nothing is made. But the theist is pretty much always an adherent of the KCA *and* LFW (it is worth looking at the meta-data for the *philpapers* philosophical survey to see evidence of this)! Indeed, as Chisholm and Kane point out, agents have to be originators of causality if LFW and ultimate responsibility are to hold. But Craig insists that this is impossible, since only God can be a prime mover, and, unmoved, move the universe.

Potential Objections

The theist seems to use one of two defences here:

1) That prior causes merely influence but do not define the decision

2) That the agent is itself a cause—and that this is fine. For example, there is the theory of agent causation supposing that agents

are different to events and event causation. People can somehow ground causal chains and decisions in a way which is different to, say, a boulder rolling down a hill, hitting a tree and a pine-cone falling out. That kind of causality has no agency.

1) can easily be answered in the following way.

One of the most common defences of Libertarian Free Will (or contra-causal free will) is what I sometimes term *the 80-20% approach*. Most people, to some degree or another, accept that our lives are at least somewhat and, in most cases, a good deal influenced. This may be by genetic, biological or environmental factors. It is hard to deny that, at the point of making a decision, we aren't having our decision influenced by external or internal motivators. This is expressed often as a claim like "Well, we are influenced quite a bit, but we still have some degree of free will" or "I think we are 80% determined, but 20% of our decision-making is freely willed". There is this idea that the will *can* override causality in some force of agency.

I will now show why this approach is entirely incoherent.

Before looking at this particular point, it is worth laying out the issue of causality and free will. Contra-causal free will is so called because it appears to want to break the rules of causality. Causality is a more difficult concept than many give credit for as you can see by this very book. Taken in its simple manner of understanding, it presents this issue, known in some parts as the Dilemma of Determinism[34]:

> …the well-known dilemma of determinism. One horn of this dilemma is the argument that if an action was caused or necessitated, then it could not have been done freely, and hence the agent is not responsible for it. The other horn is the argument that if the action was not caused, then it is inexplicable and random, and thus it cannot be attributed to the agent, and hence, again, the

agent cannot be responsible for it. In other words, if our actions are caused, then we cannot be responsible for them; if they are not caused, we cannot be responsible for them. Whether we affirm or deny necessity and determinism, it is impossible to make any coherent sense of moral freedom and responsibility.

What this means is that an action is either caused or it's random. To claim it is caused but not determined is nonsensical. The basis of rejecting LFW is this, since neither option allows for free will. To think otherwise would imply that the agent caused the action but that it wasn't determined; that the agent could have done otherwise. For example, at 9:15am this morning in this universe, when the phone rang, I could have picked it up, or I could not have done. I had the ability to do either. Or, at $t=1$ in causal circumstance CC the agent could have done X or Y. If the agent had done X and we continued the universe until $t=10$ and then rewound the universe, do you know what, he could have done Y.

The problem with this is that with CC at $t=1$, the agent had a set of reasons for doing X. This is what determined that he chose X. The universe up until that moment, his genes and biology, the environment up to every single atom, had causal influence to produce the "choice" of doing X. So if we went on 10 minutes and then rewound to CC at $t=1$, considering the entire universe would be identical, and the person identical, what *could* cause the agent to choose Y and not X? In order to do so, the agent would have to be ever so slightly different, or the environment (universe) would have to be different. In order to claim that the agent *could* have done differently would surely require a *reason*. Since a freely willed action cannot be randomly defined, as the agent has no ownership over random, then there must necessarily be a reason.

Causality takes on the form of a chain of events, when seen in its simplest (and I argue in this book, erroneous) form, but it is useful for explaining the point here. A causes B which causes C and then

D. This goes back until the beginning of the universe or some such similar causal scenario. What Libertarian free willers believe is that, since the causal chain cannot regress back to the Big Bang or similar, as this implies determinism, the agent must be the *originator* of the causal chain. Let's look at this in terms of "why" questions. Why did D happen? Because of C. Why C? Because of B, and so on. In the case of an agent, we cannot keep asking the why question because we keep going back and back, beyond the decision. So at some point, the agent has to be ultimately causally (and thus morally, it is argued) responsible for the decision—the originator of the causal chain. The problem is that without the ability to answer the why questions, the basis of the causal chain becomes "just because" which is synonymous with random or irrationality. This is why, in *Freedom Evolves*, philosopher Daniel Dennett claims that free will requires determinism since without it, there is no reason for an action and it becomes meaningless.

Which is all good and well, but what about the issue at hand? Well, when people claim we are, say, 80% determined, but that 20% of an action is still freely willed, we have *exactly* the same problem— we have just moved that argument into a smaller paradigm, into the 20%. Assuming that we forget the 80% fraction which is determined so not being of interest to the LFWer, we are left with the 20%. But this is devoid of determining reasons. So what, then, is the basis of that 20% in making the decision? The agent cannot say, "Well my genetically determined impulses urged me to A, my previous experience of this urged me towards A, but I was left with a 20% fraction which overcame these factors and made me do B" because he still needs to establish the decision as being reasonable; as having causal reasoning behind it. So if that 20% is not just random or unknown (but still grounded in something) and had any meaning, then it would be reasoned! The two horns of the Dilemma of Determinism raise their ugly heads again. We are left with reasoned actions or actions without reason, neither of which give the LFWer the moral responsibility that they are looking for.

It is not so much the scientific reasoning and evidence that the LFWer has to contend with (which is mountainous, and enough in its own right) but the metaphysical reasoning about causality which demands serious attention.

2) (that the agent is itself a cause—agent causation) is a non-starter, as far as I am concerned. Agent causation is a theory developed by philosophers like Roderick Chisholm half a century ago. I am not that sure that many people adhere to it these days, preferring models like event causation (such that people are reliant on brain states which are physical events). It seems there is no good reason for asserting that agents are causally different to standard events. One can appeal to some kind of dualism, but causality is metaphysical as a concept, and dualistic substances would surely need to "adhere" to it in the same way matter does. To merely suppose an agent can be sufficient explanation for the cause of a decision is particularly question-begging. Without causal reasons, a decision grounded in no reason other than "the agent" is synonymous with random. Brain events, genetics and biology, we know, cause agents to make the decisions they do. Mixed with the environment, and you have a causal circumstance and determinism.

We know, for example, that in the Benjamin Libet style experiments (where we can observe that the brain kicks into gear before the conscious brain "decides" to press a button), we can actually ask the subject to press a left or right button and send trans-cortical stimulation (magnetic stimulation) in to the brain and make the agent either choose left or right, depending on where we send it. The agent then assigns their own agency to that afterwards claiming that they freely chose left or right. These are just the tip of a very large iceberg which extends to the sea floor of causal determination. For further information, please refer to my book *Free Will?*[35], my article in *Free Inquiry*[36] or my chapter on free will in John Loftus's forthcoming book *Science and Christianity*[37].

I do not want to get into a debate arguing for the philosophical truth and empirical evidence in support of hard incompatibilism here, where libertarian free will is incompatible with both determinism and adequate determinism (such that quantum randomness may be apparent). No, the agent cannot be asserted as an entity able to start a causal chain, because this assumes that a causal reason is given for a causal chain, but in a causal vacuum. There can simply be no sense to be made of rational agent origination.

This all means that someone who subscribes to the KCA cannot consistently and coherently be a subscriber to libertarian free will. Theists, then, are not logically consistent.

3.5 Quantum physics

As the maxim goes, if you claim to understand quantum, then you don't understand quantum. In the context of this book, however quantum physics potentially presents quite a problem for the theist arguing for the first premise of the Kalam.

Quantum physics is concerned with what goes on in the universe at the micro-level, and the fundamental level of physical reality. When we get down to this depth, scientists use different lenses through which they interpret the data, and much of the data looks to be random, or a-causal. This is what is pertinent to the KCA. One interpretation of this quantum indeterminism is called the Copenhagen Interpretation. Under this paradigm, the causal principle appears to break down. On the other hand other interpretations are deterministic, some relying on the hidden variable idea that we just don't know enough about the data and reality to know there aren't other variables at play to cause what might otherwise appear to be random effects.

William Lane Craig together with J.P. Moreland, states:

> Nevertheless, a number of atheists, in order to avoid the argument's conclusion, have denied the first premise. Sometimes it is said that **quantum physics** furnishes an exception to premise (1), since on the subatomic level events are said to be uncaused (according to the so-called **Copenhagen interpretation**). In the same way, certain theories of cosmic origins are interpreted as showing that the whole universe could have sprung into being out of the subatomic vacuum. Thus the universe is said to be the proverbial free lunch.
>
> This objection, however, is based on misunderstandings. In the first place, not all scientists agree that subatomic events are uncaused. A great many physicists today are quite dissatisfied with the Copenhagen interpretation of subatomic physics and are exploring deterministic theories like that of David Bohm. Thus subatomic physics is not a proven exception to premise (1).

Here, then, they rely on appealing to other interpretations of quantum. This is another hoop for the adherent of the KCA to have to jump through, thus rendering the eventual probability of it being true slightly less. There are also ramifications for this strict deterministic view of causality on Craig's claims for libertarian free will, but that is another story.

It is true that where the Copenhagen Interpretation once held sway over scientists, it has, in recent years, become slightly less prevalent; yet it is still a major player in the range of interpretations that exist. It should be noted that while many contemporary physicists are abandoning the Copenhagen interpretation for more deterministic interpretations, these other interpretations are not friendly to the KCA. In fact the most popular competitor to the

Copenhagen interpretation is the Many Worlds interpretation, which supports the idea of a multiverse. This interpretation would cause an entirely separate set of problems for the KCA, but they would be no less fatal.

Craig, though, continues by levelling a second counterpoint at this line of argument:

> Second, even on the traditional, indeterministic interpretation, particles do not come into being out of nothing. They arise as spontaneous fluctuations of the energy contained in the subatomic vacuum, which constitutes an indeterministic cause of their origination.

On the other hand, cosmologist Alexander Vilenkin disagrees[38]:

> If there was nothing before the universe popped out, then what could have caused the tunneling? Remarkably, the answer is that no cause is required. In classical physics, causality dictates what happens from one moment to the next, but in quantum mechanics the behavior of physical objects is inherently unpredictable and some quantum processes have no cause at all.

If Craig is stating that the Copenhagen Interpretation *might* be false, then this means the strength of his premise is equally affected: Everything that begins to exist *might* have a cause for its existence.

Matthew Denigris, of MIT, states more forcefully (in his paper "On the Kalam Cosmological Argument")[39]:

> If these questions are indeed problematic (meaning they are fundamentally incommensurable with Aristotelian causation), we can take only one of the following two positions: either the Copenhagen interpretation is completely wrong or our intuitive ideas

regarding causation need to be modified or dispensed with. Even if philosophers of quantum mechanics can reconcile these questions with strict Aristotelian causality, there remains an ever-present burden of proof on those who assert that traditional causal relationships continue to hold beyond subsets of the universe, at the cosmological level, and are a more-legitimate way of interpreting the double-slit experiment in terms of explanatory power, parsimony, or otherwise.

It is worth taking a moment to address a possible misinterpretation of my argument. I am not at all claiming that causality is rendered incoherent or nonexistent by our explorations into quantum mechanics and the early universe. In fact, I am personally disposed to the position that causal mechanisms (of varying forms) are a fundamental characteristic of reality as we know it. Rather, my claim is that because causality manifests itself in ways that are not intuitively obvious to us (i.e. in ways that seemingly violate Aristotelian causality), we have no good reason to believe that causal relationships either (i) exist at the cosmological level or, a weaker claim, (ii) are generalizable as a metaphysical principle - which by definition extend beyond the universe - from our experience within the universe. My consideration of the dynamics of the early universe and of the implications of the Copenhagen interpretation of quantum mechanics serves principally to evidence the foundations of these claims. Therefore, the generalization of causality implied by the conclusion of the KCA (which I have incorporated into my analysis of P1 in general) rests upon a weak foundation. The failure of Aristotelian causality within subsets of the universe renders Craig's cosmological extrapolation at best uncertain. If we again consider the fact that Craig's argument relies upon creation ex nihilo, a process that is wholly unfounded in any empirical evidence and not extrapolatable from any principles that we are currently familiar with, it is immediately clear that P1 is untenable.

So Craig *needs* the Copenhagen Interpretation to be wrong, and, what's more, I don't think his second objection gets him out of jail free. It seems like, in appealing to a quantum vacuum, Craig is appealing to there being *something* at root to eventually cause, in a generalised manner, these quantum events; however, he is not looking at these quantum events in and of themselves. Craig is narrowing his approach down to specifically material causation. This is potentially a slippery move that attempts to hide what is a rather weak counterpoint. As Denigris continues: "the existence of viable alternatives to our traditional notions of causality within the universe at the very least calls into question P1's obviousness or necessity, especially when applied to the universe itself". If Craig is taking universal causality as an axiomatic intuition, he seems to want to cherry pick certain types of causality to fit his agenda.

There are other proposals, concerning quantum physics, which present further problems for Craig. Virtual particles, which appear and disappear, apparently at random, from observation have been suggested to counter such claimed universal causality. Similar claims have been made to suggest that the universe could start in such a manner. Craig resorts to the previously mentioned objection that such quantum events aren't producing new energy[40]:

> For virtual particles do not literally come into existence spontaneously out of nothing. Rather the energy locked up in a vacuum fluctuates spontaneously in such a way as to convert into evanescent particles that return almost immediately to the vacuum.

Now Craig is playing into my earlier objection to the KCA: that the universe is a set of existing matter and energy that acts as one single example of causality. If Craig is happy to defer all other causal events "within" the universe back to an earlier material cause (here, a quantum vacuum), then we are back to making an (intuitive) assertion into a rule derived from a single occurrence. The quantum vacuum produces everything else, so what produces the quantum vacuum? Well, he can't really employ an intuition if causality derived from observations within the universe because he has special pleaded examples of causality therein as not being applicable. Apparent random events don't qualify as being a-causal and defying the first premise because there was a material cause for them (i.e. the quantum vacuum). So we can only really look at the material cause, which means you cannot employ an intuition derived from causes and effects (and *not all of them* to boot) further down the chain (or matrix) for fear of double standards and special pleading. The resultant intuition, then, is empty and baseless.

PART FOUR
Premise 2

The second premise "The universe began to exist" is also victim to a number of objections, both similar and different to the objections to premise 1. Obviously, premise 2 is dealing with the universe and attaching the behaviour of everything (else) and claiming "the universe" as part of the subset of "everything". Craig, for example, uses four approaches to the second premise, two of which are *a priori* and two of which are *a posteriori*. This means that two defences (a priori) of premise 2 are self-evident, proceeding from theoretical deduction; or in other words, deductive in quality. The other two defences (a posteriori) depend on observations and experiences, thus being inductive in quality.

The two a priori defences regard infinites in claiming that if one asserts that the universe is eternal, infinitely going into the past, then this is philosophically and mathematically incoherent, and, therefore, the universe necessitated a start. To defend the premise inductively, Craig appeals to modern cosmology such as the Big Bang theory (or certain models thereof) and the second law of thermodynamics.

4.1.1 Can the universe have existed infinitely into the past?

Although there is an intuitive element to what Craig states, this is something worth looking at. Craig claims that the universe must have begun to exist because the notion that it existed infinitely into the past would mean that we would never get to "now", in simplistic terms. Craig has set out the argument as follows[41]:

2. The universe began to exist.

2.1 Argument based on the impossibility of an actual infinite.

> 2.11 An actual infinite cannot exist.
> 2.12 An infinite temporal regress of events is an actual infinite.
> 2.13 Therefore, an infinite temporal regress of events cannot exist.

2.2 Argument based on the impossibility of the formation of an actual infinite by successive addition.

> 2.21 A collection formed by successive addition cannot be actually infinite.
> 2.22 The temporal series of past events is a collection formed by successive addition.
> 2.23 Therefore, the temporal series of past events cannot be actually infinite.

It seems, when we look at the work of Craig and James Sinclair in their chapter on the Kalam and infinities in the *Blackwell Companion to Natural Theology*[42], that on *face value* actual infinites do not exist. There is a difference between an actual infinite (say, a collection of things) and a mathematical (abstract or potential) infinite which can be used within mathematical calculations and theorems, but doesn't entail an actual, real set of objects existing. Craig contests the existence of *actual* infinites. This is crucial for his argument.

It is worth noting that Christian postdoctoral researcher Aron Wall, who at the time of writing is studying quantum gravity and black hole thermodynamics at UC Santa Barbara, isn't quite in agreement with fellow Christian Craig, here. In a piece written about open and closed universes, he states that he is in agreement with the statement: the logical argument for an impossibility of an actually infinite past is invalid, as an actually infinite past is not logically impossible. He states:

> t=−∞ is not a time, rather it is the limit of a sequence of arbitrarily early times. There are (not necessarily) any actual "objects" or "things" existing at t=−∞, rather things exist at finite values of time. So I would answer the objection you raise by saying that no actual entities ever existed which would still need to "traverse" an infinite amount of time. (Just as objects can exist at arbitrarily late moments, in the other time direction.) No two actually existing times differ by a[n] infinite amount of duration....
>
> I personally think models that begin with a finite past are more elegant. But lack of elegance is not the same thing as logical impossibility.

What the basic argument that Craig and others posit is this: if there have been infinitely many days before today, then there must be some day in the past such that there are infinitely many days between that day and today. But does this follow? Let me pull on the expertise of my mathematical friend, Dr. James A. Lindsay[43], in saying the following. It is well worth reading his superb book on infinity and God, entitled *Dot, Dot, Dot: Infinity Plus God Equals Folly.*[44]

Let's put this into a simple example. Once, in the past, there was a moment. Let's call it E. Time then passed, moment after moment, until we reached P, the present. If this time difference is infinite, then surely we have a problem traversing that infinite to the present, to P?

The distance between any two points on an infinite number line is still always finite. Therefore, the given construction, one of E or P, is impossible, or the space between them is finite. There is much to be said here.

So imagine that there was some event E infinitely long ago and some time P at the present moment. We will suppose something happened at time E that we might be interested in, say a star went supernova—something we can see that emits light (which has a finite speed). Now imagine the light travelling out from E. It's coming towards us, and after some time, it is halfway to us. If we can think of E and P both being points on a line, then when does this halfway happen? Only after infinite time passes because half infinity is still infinity. That is, it never gets here (or to half of here, or to half of that, or to half of that, or to half of that,...).

What's the difference between an event E that cannot possibly be known by any means to have happened and an event E that never happened? Hint: nothing.

What we do is look back from P towards E (into the past), and we imagine that we can look back to it. Conceptually, just a little more back, and we're there. But we're wrong. We're always wrong. Our notation "..." tricks us. It's easy to write E...P, but those three dots cannot be shorthand for infinity there.

In fact, mathematically, if we have $x...y$ and it is always defined that there are $y-x-1$ values between x and y, it can never be infinity.

If we wanted to denote infinity, we would say something like x, $x+1$,... (and leave the ellipsis open-ended). In other words, we're lying to ourselves when we say "imagine some event E that happened infinitely long ago, like this E........P, where P is the present moment." No such thing exists.

If we look from E toward P, we only ever see E+(however long we've looked). That we never see anything like P is the definition of infinity.

Craig, though, is just slipping an event E in the back door. He changes "the universe began finitely long ago or it didn't" to "the universe began finitely long ago or it began infinitely long ago", which isn't the same because it assumes "began".

"Eternal" means that Craig's whole construction (counting up from E to show that it will never get to P) is erroneous because E can't be said to exist in the first place.

This is quite a conceptual shift to grasp. Think of a number line. How many values are bigger than P, for any value P? Less than? In both cases, you would have infinity.

If you had to start at the left end (a beginning) and travel to the right, you'd never get to zero (or any value at all, actually). If you're on the number line anywhere, then by definition you didn't start at an end. You started somewhere "in the middle", with an infinite number of values smaller and an infinite number larger.

Since we're at time P, we are somewhere on the timeline. That means, if the universe is eternal, there are infinitely many past moments and infinitely many future moments. Later, at time Q, it's the same. We can measure the number of moments that passed between P and Q, and that amount is always finite, but we cannot place any event (E or Q) "infinitely" far away. But this doesn't contradict the infinite length of the line.

That's why Craig's argument depends upon assuming that there either was a beginning finitely long ago or infinitely long ago. The second he says, "It cannot be infinitely long ago because an infinite number of moments cannot have passed to get to now", he's assuming (for the sake of absurdity) that there is some point E negative infinity moments ago. No such point exists on an eternal time line. In other words, he's referencing something that doesn't exist.

If the universe is eternal, then no matter what time you designate as P, there are infinitely many moments before it and infinitely many after it. There is no external reference point from which to get to P. Things are just happening at P because things have literally always been happening and literally always will be.

That's what eternal means. Every point in time, literally all of them, is in the middle of the action. I think it reduces to having to say the universe is a brute fact, but I can't say there is a more sensible choice for a brute fact than the universe. Indeed, you can argue that this is what Craig does anyway when he says God is a brute fact (or necessary, or infinite).

The universe could be defined as "everything that is, will be, or ever was; all of existence; existence itself". I think that qualifies as a candidate for a legitimate brute fact.

Dr. James East, a friend of mine and a mathematician, recently wrote a paper for the journal *Faith and Philosophy*. The journal has kindly given permission for him to add that contribution to this book. The paper, "Infinity Minus Infinity"[45], looks at the arguments Craig uses in concluding that infinites cannot actually exist in order to show that a past infinite universe cannot be possible, and finds them to be wanting. Here is his paper in its entirety. Actual infinities may indeed be impossible, or may exist in some world or possible world, but Craig's arguments cannot show that they do not.

Before embarking on Dr. James East's section, it is worth giving a small glossary of terms for reference to the reader. There are also some useful endnotes for this section:

Cardinal: The cardinality of a set is the number of elements it has. A cardinal is the cardinality of some set. These can be finite or infinite. For example, the cardinality of the set of people in China is roughly one billion. The cardinality of the set of integers (whole numbers) is infinite.

Transfinite: A cardinal is transfinite if it is bigger than all finite cardinals. The cardinals of the set of all integers and the set of all real numbers are both transfinite, though these cardinals are not equal. There is a never-ending supply of transfinite cardinals.

Countable: The smallest transfinite cardinal is that of the natural numbers $N = \{0,1,2,3,...\}$. A set is called countable if it has the same cardinal as N (i.e., the elements of the set may be matched up to the elements of N). The cardinal of the set N is denoted $\aleph_0$.

א (aleph): The first letter of the Hebrew alphabet denotes numbers representing the sizes or cardinalities of infinites. $\aleph_0$ is the smallest set, representing the size of the set of natural numbers (N, as seen above)

4.1.2 Infinity Minus Infinity, by Dr. James East

In this section, I consider an argument advanced by William Lane Craig and James D. Sinclair against the possibility of actual infinite collections based on Hilbert's Hotel and alleged problems with inverse operations in transfinite arithmetic. I aim to show that this argument is misguided, since it is based on a mistaken view that the impossibility of defining $\aleph_0 - \aleph_0$ [the smallest sized infinite set subtracted from the smallest sized infinite set] entails the impossibility of removing an infinite subcollection from an infinite collection.

1. Introduction

Kalām Cosmological Arguments seek to establish the existence of a First Cause based on the premise that the universe had a beginning. This premise is typically supported either by empirical data (such as Big Bang cosmology) or by an argument against the possibility of an infinite past. In the *Blackwell Companion to Natural Theology* (2009), William Lane Craig and James D. Sinclair argue against the possibility of an infinite past as follows:

> P1. "An actual infinite cannot exist.
> P2. An infinite temporal regress of events is an actual infinite.

C. Therefore, an infinite temporal regress of events cannot exist."[46]

They offer several supporting arguments for both of the premises, and this note concerns the supporting argument for P1 based on Hilbert's Hotel and an alleged problem with inverse operations in transfinite arithmetic. According to Craig, the "strongest arguments in favour of the impossibility of the existence of an actual infinite" are "those based on inverse operations performed with transfinite numbers".[47] Craig continues to favour this style of argumentation in his public lectures and debates, many of which are readily available online.

The familiar story of Hilbert's Hotel involves a hotel with infinitely many rooms, numbered 1,2,3,..., each of which is occupied by a guest. When a new guest arrives, the proprietor asks the guest in Room N to move to Room N + 1 (for N = 1,2,3,...), thereby freeing up Room 1 for the new guest to use. The proprietor can even accommodate (countably) infinitely many new guests by asking the guest in Room N to move to Room 2N, thus freeing up the odd numbered rooms for the new guests. But, according to Craig and Sinclair, "Hilbert's Hotel is even stranger than the German mathematician made it out to be"[48]. Indeed, as they observe, if all the guests in odd numbered rooms (1,3,5,...) check out, there will still be infinitely many guests remaining: all those in even numbered rooms (2,4,6,...). However, if all the guests in Rooms 4,5,6,... checked out, then the hotel would be nearly empty, with only three rooms remaining occupied.

The two scenarios above indicate that one could remove infinitely many objects from an infinite collection (if one existed) in two different ways, and end up with a different number of objects left over. And this, Craig and Sinclair allege, is "absurd... Can anyone believe that such a hotel could exist in reality?"[49] Craig and Sinclair rightly note that "inverse operations of subtraction and division with infinite quantities are prohibited" in transfinite arithmetic, but protest that "in reality, one cannot stop people from checking out of a hotel if they so desire!"[50] In other words, they allege that there is some kind of disconnect between mathematics and reality, in that mathematical considerations somehow imply the impossibility of an action (checking out of a hotel) that we know should be possible (no matter how many rooms there are).

I do not claim to know that Hilbert's Hotel could exist in some metaphysically possible version of reality, but I aim to show that the reasons Craig and Sinclair give for rejecting the possibility are flawed. This will involve recalling the two standard methods of defining subtraction for finite quantities: as the inverse operation of addition, and via the "taking away" operation. I will explain why both methods lead to the impossibility of defining $\mu - \mu$ where μ is an infinite cardinal, but not to the impossibility of performing certain tasks such as checking out of an infinite hotel.

2.   Subtraction as the inverse operation of addition

When (finite) arithmetic is taught to young children, addition is usually the first concept taught. Students are asked to complete exercises such as $2 + 5 = \square$. As they become more advanced, they move on to exercises like $3 + \square = 9$. Since the solution to the equation $a + \square = b$ (with a and b finite) is uniquely determined by a and b, we are able to define $b - a$ as "the unique solution to the equation $a + \square = b$".

The above considerations of guests arriving at Hilbert's Hotel indicate that the equations $\aleph_0 + 1 = \aleph_0$ and $\aleph_0 + \aleph_0 = \aleph_0$ both hold. (Here, $\aleph_0$ denotes the cardinality of the set $\{1,2,3,..\}$ of all natural numbers.) This shows that the equation $\aleph_0 + \square = \aleph_0$ does not have a unique solution; indeed, it has infinitely many solutions. For this reason, we are unable to define the difference $\aleph_0 - \aleph_0$ as "the unique solution to the equation $\aleph_0 + \square = \aleph_0$". To put it differently, knowing that $\aleph_0$ was added to an unknown quantity to give a total of $\aleph_0$ is not enough information to deduce the value of the unknown quantity. This means that the operation of adding $\aleph_0$ is not invertible.

But this does not entail that actual infinite collections are impossible. Many real world phenomena are modelled by mathematical operations that are not invertible: squaring numbers, multiplying matrices and composing functions are all examples. Since this does not give us cause to doubt the existence of such real world phenomena, neither should we reject the possibility of an actual infinite collection simply because such collections would be modelled by non-invertible mathematical operations.

But, as mentioned above, Craig and Sinclair argue that the most severe problems concern the situation encountered when guests check out of the hotel. For does this not seem to be a case in which we are trying to define $\aleph_0 - \aleph_0$? This leads us to consider the second way to define subtraction.

3. Subtraction as taking away

Recall that subtraction (of finite quantities) may also be defined without explicit reference to addition. One way to do this is to make use of the "taking away" operation. In fact, sometimes "5 − 3" is read as "five take away three". To help a child calculate 5 − 3, a teacher will often say something like:

"If you had 5 apples, and I took away 3 of them, how many would you have left?"

Repeated experimentation shows that it does not matter which three apples are removed: there will always be two left. You could also perform the experiment with bananas rather than apples, and the answer will always be the same. If you start with any collection of 5 objects, and remove any 3 of them, you will always end up with 2. Because of this, we can define 5 − 3 to be "the number of objects left when you take any 3 objects away from any collection of 5 objects". These considerations are just special cases of the following basic theorems from set theory.

Theorem 1. Suppose A is a finite set. Suppose $B \subseteq A$ and $C \subseteq A$, and that $|B| = |C|$. Then $|A \backslash B| = |A \backslash C|$.[51]

Theorem 2. Suppose A and B are finite sets and that $|A| = |B|$. Suppose $C \subseteq A$ and $D \subseteq B$, and that $|C| = |D|$. Then $|A \backslash C| = |B \backslash D|$.

But note that Theorems 1 and 2 are stated in terms of finite sets. Theorem 1 cannot be proven in the absence of the assumption that A is finite. Likewise, the finiteness of A and B is essential in proving Theorem 2. In fact, the theorems become false statements if we remove the word "finite", for we also have the following theorem.

Theorem 3. Suppose A is a countably infinite set. Then there exist subsets $B \subseteq A$ and $C \subseteq A$ such that $|B| = |C|$ but $|A \backslash B| \neq |A \backslash C|$.[52]

Theorem 3 shows that it is not possible to define $\aleph_0 - \aleph_0$ as "the number of objects left when you take any $\aleph_0$ objects away from any collection of $\aleph_0$ objects".[53] The number of objects left after one removes $\aleph_0$ objects from a collection of $\aleph_0$ objects will depend on which objects were removed.

But, again, this does not entail that actual infinite collections are impossible. And neither does it entail that one could not remove an infinite number of objects from an infinite collection, if one existed. If the proprietor of an infinite hotel told you that infinitely many guests had just checked out, this information alone would not allow you to determine how many guests remained; the number of guests remaining would depend on *which guests* checked out. And this is very different from saying that the guests *could not* have checked out.

4. Where is the contradiction?

It is possible that Craig and Sinclair anticipated something like the case I made in the previous section. In a footnote, they say[54]:

> It will not do, in order to avoid the contradiction, to assert that there is nothing in transfinite arithmetic that forbids using set difference to form sets. Indeed, the thought experiment assumes that we can do such a thing. Removing all the guests in the odd-numbered rooms always leaves an infinite number of guests remaining, and removing all the guests in rooms numbered greater than [three] always leaves three guests remaining. That does not change the fact that in such cases identical quantities minus identical quantities yields nonidentical quantities, a contradiction.

Elsewhere, they say that[55]

> ...the contradiction lies in the fact that one can subtract equal quantities from equal quantities and arrive at different answers.

So it seems that Craig and Sinclair think that the scenario illustrated in the story of Hilbert's Hotel contradicts a principle like:

(i) Removing identical quantities from identical quantities yields identical quantities.

But what reason do we have to accept this principle? Notice the similarities to Theorem 2 which, stated in similar terms, says:

(ii) Removing identical quantities from identical *finite* quantities yields identical quantities.

We know that Principle (ii) can be proved mathematically. But in order to extend Principle (ii) to the stronger Principle (i), one would need to provide an argument in favour of the principle:

(iii) Removing identical quantities from identical *infinite* quantities yields identical quantities.

Note that acceptance of Principle (iii) does not require a commitment to the existence of actual infinite quantities; indeed, it is a vacuous statement if there are no actual infinite quantities. Since we have seen (in Theorem 3) that Principle (iii) would be false if infinite collections do exist, a proof of Principle (iii) (and, hence, of Principle (i)) would require a proof that there are no infinite collections. And, of course, if the goal is to use Principle (i) to prove that there are no infinite collections, then this would render the entire argument circular.

5. Conclusion

If actual infinite collections were to exist, then they would naturally have properties that were not shared by finite collections. For one obvious example, if one attempted to count through an actual infinite collection at a constant pace, then one would never finish (and this is also the case with potential infinite collections, such as a future eternity of discrete days). The story of Hilbert's Hotel simply highlights another such property that distinguishes actual infinite collections from finite ones: just knowing that an infinite subcollection has been removed from an infinite collection of objects does not allow one to determine how many objects remain. But this property itself does not entail that actual infinite collections are impossible.

James East
Centre for Research in Mathematics
School of Computing, Engineering and Mathematics
University of Western Sydney

What East has done here is to show, rather like I do here with the thesis as a whole, that whilst an actual infinite may not exist, you cannot show that using the methods that Craig relies on. In the same way, the universe *may* have had a beginning and that *may* have been caused by a god, but you cannot show that using the Kalam Cosmological Argument.

4.2 The premise as inductive

Aside from this notion of infinity, and as with the first premise, this second premise is victim to being an assertion made on the basis of inductive observations. That the universe began to exist is an assertion derived from scientific research and observations which is, by definition, a posteriori. As mentioned above, whether or not the idea that the universe was past eternal holds, the defence of premise 2 remains purely inductive. Which again infers that the deductive nature of the KCA rests upon inductively derived premises. This has the same effect as discussed earlier, making the argument only as powerful as the inductive strength of the propositions and evidence used to negotiate the conclusions of the inductive premises (in this case, that "the universe began to exist").

Given the nature of physics, and the ever changing landscape of modern cosmology, it is a risky business to put all one's eggs into a single theoretical basket, as we shall later see.

4.3 Creation of the universe ex nihilo

As mentioned earlier in the context of the quote from Wes Morriston, the idea that the universe itself falls victim to "everything that begins to exist has a cause" is an unevidenced assertion. I would like to take the opportunity to expand more broadly upon this point. The assumption by philosophers like Craig is that the universe must act in accordance with the behaviour of causality of discrete objects within the universe itself. Whilst I have shown this to be incoherent for several reasons, let me assume that the problems with causality and abstracta are non-existent. Let us assume that there are discrete objects within the universe which require causality for their existence. Can this behaviour be projected on to the universe itself? Morriston disagrees that it *necessarily* can be. To repeat[56]:

> Now contrast the situation with regard to the beginning of time and the universe. There is no familiar law-governed context for it, precisely because there is nothing (read, "there is not anything") prior to such a beginning. We have no experience of the origin of worlds to tell us that *worlds* don't come into existence like that. We don't even have experience of the coming into being of anything remotely analogous to the "initial singularity" that figures in the big bang theory of the origin of the universe.

I must clarify myself here because earlier I talked as the universe not being separate from anything *in* the universe, as the universe *is* everything. Thus I reject other criticisms of the KCA as falling foul of the *fallacy of composition*. This is the inference that something which is true of the parts must be true of the whole. Examples might be:
- Hydrogen is not wet, oxygen is not wet, and therefore water (H_2O) is not wet.

- Cells weigh a minute amount. I am made up of cells. Therefore, I weigh a minute amount.

People do try to claim adherents of the KCA commit this fallacy, but I deny this because the whole simply *is* the parts. The term "universe" is merely a term to refer to the collection of those parts. Water, for example, is made up of those parts which make a different "thing" (water) than the parts. The universe, other than the human abstract label to refer to everything, is no separate "thing" from "everything" which constitutes it.

Craig is very much relying on the medieval maxim *ex nihilo nihilo fit*—out of nothing, nothing comes (nothing comes from nothing). This is the intuition upon which Craig rests so much authority. But as a maxim, it is only as good as its inductive nature (i.e. it is not *necessarily* true), and those observations were made of discrete objects within the framework of space and time provided by, or which constitutes part of, the universe itself. There is absolutely no guarantee that such a maxim applies to non-spatiotemporal dimensions (if such a term is even possible).

In other words, within our universe, we can see that nothing just pops into existence (and we will park quantum tunnelling, entanglement and any other number of scientifically weird phenomena which may disprove this maxim). However, we cannot take that observation, and apply it to without the universe, because that is a *false analogy* or *non sequitur*.

I would also tend towards the notion that, even given God's supposed omnipotence, it is still incoherent that God could produce anything *ex nihilo*. There is something inherently, and intuitively, problematic about the whole process of creation *ex nihilo* that I find hard to swallow. Though this appears to be mere intuition on my part, Adolf Grünbaum, writing in the journal *Philosophy of Science*, masterfully sums up this issue[57]:

> Therefore, if creation out of nothing (ex nihilo) is
> beyond human understanding, then the hypothesis that

it occurred cannot explain anything. Even less can it then be required to fill explanatory gaps that exist in scientific theories of cosmogony [dealing with the origin of the universe]. Indeed, it seems to me that if something literally passes all understanding, then nothing at all can be said or thought about it by humans. As Wittgenstein said: Whereof one cannot speak, thereof one must be silent. Dogs, for example, do not bark about relativity theory. Thus, any supposed hypothesis that literally passes all understanding is simply meaningless to us, and it certainly should not inspire a feeling of awe. To stand in awe before an admittedly incomprehensible hypothesis is to exhibit a totally misplaced sense of intellectual humility! It is useless to reply to this conclusion by saying that the creation hypothesis may be intelligible to "higher beings" than ourselves, if there are such. After all, it is being offered to us as a causal explanation!

However, as philosopher Wes Morriston states in his essay *Creation Ex Nihilo and the Big Bang*, given that the "Greatest Conceivable Being" would be one that could create *ex nihilo*, the idea of whether God can possibly create anything *ex nihilo* becomes an argument over conceivability rather like the Ontological Argument. (For those who do not know of this argument, it essentially argues that God, being defined as most great or perfect, must exist, since a God who exists is greater than a God who does not. It essentially magics God into existence through what some say is a linguistic trick.)

Craig, though, confirms his belief in a universe starting out of nothing as follows[58]:

> This event that marked the beginning of the universe becomes all the more amazing when one reflects on the

69

fact that a state of "infinite density" is synonymous to "nothing." There can be no object that possesses infinite density, for if it had any size at all it could still be even more dense. Therefore, as Cambridge astronomer Fred Hoyle points out, the Big Bang Theory requires the creation of matter from nothing. This is because as one goes back in time, one reaches a point at which, in Hoyle's words, the universe was "shrunk down to nothing at all." Thus, what the Big Bang model of the universe seems to require is that the universe began to exist and was created out of nothing.

This argument can be boiled down to making something of "infinite density" synonymous with "nothing". Because no object can have infinite density, it seems, Craig feels justified in having this conclusion. It is interesting that Craig accuses cosmologists like Laurence Krauss, with reference to his book *A Universe from Nothing*, of equivocating over the term "nothing" such that it actually means "something" (such as a vacuum fluctuation model)[59] and yet is seemingly allowing such a concept traction here. As Morriston responds[60]:

> "Infinite density" is not synonymous with "nothing", and the "initial singularity" that figures in the big bang theory is not simply nothing at all. A mere *nothing* could not explode, as the infinitely dense particle is supposed to have done. And even if it lacks spatial and temporal spread, the initial singularity has other properties. For starters, it has the property of "being infinitely dense." It is therefore a quite remarkable *something*, and not a mere nothing.

This is crucial since it seems that Craig is adopting double standards in accusing Krauss of misrepresenting ideas of "nothing" whilst at the same time holding a position which claims that something with at least the properties of infinite density can also be known as "nothing". How can something (for want of a more neutral word) with the positive property of infinite density also be nothing at all? Either Craig's point is incoherent itself, or the very notion that there can be an object with infinite density is false, thereby invalidating the Big Bang theory and Craig's contingent defence. As Morriston continues, "it appears that the big bang model of the origin of our universe provides no support for the claim that the universe was created *out of* nothing."

This time at the singularity, where there is infinite density, is known as the Planck Epoch (or Planck Wall) where the known physics breaks down. As Craig claims, "What makes their proof so powerful is that it holds regardless of the physical description of the universe prior to the Planck time."[61] By Planck time, I surmise he means the Planck Epoch. Craig comes across the same problem that was discussed above in that this version of existence at this time cannot be nothing in the sense he claims.

The fact remains that it is no more coherent that a personal, non-temporal entity with all the associated attributes, known as God, is any more likely to have existed "prior" to the Big Bang in order to have caused all the matter and energy to pop into being out of nothing, than some other cause. By Ockham's Razor, we are already confusing matters with an entirely new layer of unexplained entity which ends up being a divine brute fact. The postulation that the universe is a brute fact (whether eternal, oscillating, one-off, multiverse or otherwise) is simpler and thus more attractive.

Or:

1)   The universe as a brute fact

Is simpler and more attractive than

2)   The universe + God as a brute fact

We could get on to talk about necessary versus contingent entities. But if one can claim that God is necessary, in a philosophical sense, I cannot see why matter cannot also be. There just has to be something, in all possible worlds, whether God or matter.

The God hypothesis is further confused here by apparently making this contingent world of ours necessary. If God, under classical theism, has necessarily the attributes he has, and these necessarily lead to the most loving (etc.) decisions, and that this world is necessarily borne out of that, then this world is, in some sense, necessary.

But that's another conversation for another time.

4.4 Scientific theories to explain the universe and everything

In order for Craig, and other authors of the KCA, to uphold the argument's second premise (the universe began to exist), they must at least establish:

1) An initial singularity took place such that the universe began to exist.
2) That it did so out of nothing and with no preceding cause (energy, matter, force etc.) other than God.
3) That an eternal universe is impossible.
4) That an oscillating universe is impossible.
5) That the multiverse theory is either impossible or requires a similar sort of beginning/singularity.
6) That a cyclical model for the universe is impossible.
7) That "no-boundary" models are impossible or require a beginning/singularity.
8) And so on. That no other model for the universe is remotely possible.

Now, this is obviously not a cosmological book in the scientific sense, and I have neither the required expertise nor the space and time (pun intended) to make it one. Therefore, my objections will come from a philosophical foundation, though I will venture into aspects of cosmology where necessary. In fact, I might as well start now. Craig posits the Borde-Guth-Vilenkin (BGV) theory of inflation to suppose that the universe had a finite beginning. As Craig states, "the implication of the Borde-Guth-Vilenkin theorem is that the universe and even the multiverse, should there be such a thing, had an absolute beginning. Therefore, we have good grounds for thinking that the cause is not physical."[62] As he set out in *A Reasonable Faith:*

> In 1994, however, Arvind Borde and Alexander Vilenkin showed that any spacetime eternally inflating toward the future cannot be "geodesically complete" in the past, that is to say, there must have existed at some point in the indefinite past an initial singularity. Hence, the multiverse scenario cannot be past eternal...

> In 2003 Borde and Vilenkin in cooperation with Alan Guth were able to strengthen their conclusion by crafting a new theorem independent of the assumption of the so-called "weak energy condition," which partisans of past-eternal inflation might have denied in an effort to save their theory. The new theorem, in Vilenkin's words, "appears to close that door completely." Inflationary models, like their predecessors, thus failed to avert the beginning predicted by the Standard Model.[63]

At best, Craig is setting up a conclusion that is more probable than alternative theories. Science, and in particular the nascent field of cosmology, is not a discipline that welcomes indubitable assertions of fact. Craig appeals to the notion that the BGV theory is generally accepted by physicists. It does not follow, however, that the theory is not without its problems or that there are not other competent theories which provide interesting alternatives to the initial singularity which is *entirely necessary* for the KCA to work. Not only is the KCA dependent upon a set of assumptions and assertions, as seen earlier, for premise 1 to work, but premise 2 also relies upon certain assumptions; primarily, that the BGV is true and that there was, indeed, an initial singularity that implied a definite beginning to the universe (or multiverse).

As particle physicist and author Victor Stenger replies in *The Fallacy of Fine-Tuning: Why the Universe is not Designed for Us.*[64]:

> [The Borde-Guth-Vilenkin theorem] has been used by William Lane Craig to argue that the universe itself had to have a beginning. We saw that cosmologists I contacted, including Vilenkin, Carroll, and Aguirre, all of whom have published works on the subject, agreed that no such conclusion is warranted.

Sean Carroll, renowned theoretical physicist who has debated (and in my opinion beaten) Craig, has stated that[65]:

> ...the correct thing to say about the Big Bang is not that there was no time before it, it is that our current understanding of the laws of physics gives out at that moment in time. We need to think a little harder. We should be open minded: it could have been the beginning; or it could have been a phase through which the universe goes.

The singularity, where time and general relativity break down, signalling the beginning to the universe is more accurately interpreted as being merely the beginning to inflation. If we look at the idea of this singularity that Craig maintains started the universe, Stephen Hawking, in *The Grand Design*, declares[66]:

> Although one can think of the big bang picture as a valid description of early times, it is wrong to take the big bang literally, that is, to think of Einstein's theory [general relativity] as providing a true picture of the origin of the universe. That is because general relativity predicts there to be a point in time at which the temperature, density, and curvature of the universe are all infinite, a situation mathematicians call a singularity. To a physicist this means that Einstein's theory breaks down at that point and therefore cannot be used to predict how the universe began, only how it evolved afterward.

Part of the problem may also be a potential equivocation over the term "universe" as employed by Craig and the people upon whom he relies. Craig (2007) defines "universe" as "the whole of material reality" whereas Alexander Vilenkin himself defines it as:

> It is certainly more than what we can have access to. Regions beyond our cosmic horizon are included. But if there are other universes whose space and time are completely disconnected from ours, those are not included. So, by "universe" I mean the entire connected spacetime region.[67]

Of course, it isn't just the BGV which leads Craig and others to defend the KCA[68], but the theorem acts as a major supporting joist. Without the BGV, Craig's claims of the KCA are a good deal hollower.

It is important to note that most physicists appear to agree that singularities point to nothing more than incomplete knowledge or theories as Curiel and Bokulich state in "Singularities and Black Holes":

> Indeed, in most scientific arenas, singular behavior is viewed as an indication that the theory being used is deficient. It is therefore common to claim that general relativity, in predicting that spacetime is singular, is predicting its own demise, and that classical descriptions of space and time break down at black hole singularities and at the Big Bang. Such a view seems to deny that singularities are real features of the actual world, and to assert that they are instead merely artifices of our current (flawed) physical theories. A more fundamental theory—presumably a full theory of quantum gravity—will be free of such singular behavior.

In other words, if you are relying on singularities to get you to the conclusions (Premise 2) that the universe began to exist, then you are misguided and relying on incomplete knowledge.

Craig is a massive fan of the BGV because he sees it as clearly evidencing his argument. Vilenkin, himself, is less certain[69]:

> Theologians have often welcomed any evidence for the beginning of the universe, regarding it as evidence for the existence of God ... So what do we make of a proof that the beginning is unavoidable? Is it a proof of the existence of God? This view would be far too simplistic. Anyone who attempts to understand the origin of the universe should be prepared to address its logical paradoxes. In this regard, the theorem that I proved with my colleagues does not give much of an advantage to the theologian over the scientist.

Furthermore, Vilenkin spoke to Victor Stenger by email to state in reply to the question as to whether the BGV required a beginning to the universe[70]:

> "No. But it proves that the expansion of the universe must have had a beginning. You can evade the theorem by postulating that the universe was contracting prior to some time."

To which Stenger concludes:

> This is exactly what a number of existing models for the uncreated origin of our universe do.

Let me re-emphasise this. The theory does not show that the universe had a beginning, but that the expansion of the universe had a beginning. This is a vitally important distinction.

What is interesting is that none of the three members of the BGV triumvirate are theistic. Their theory has not led them to conclude what Craig has co-opted it to evidence. Indeed, Vilenkin argues for the multiverse[71], which defeats Craig's fine-tuning argument, and says the universe can have no cause[72]. Alan Guth states, or rather understates that: "It looks to me that probably the universe had a beginning, but I would not want to place a large bet on the issue."[73] He also says that "Conceivably, everything can be created from nothing. And 'everything' might include a lot more than what we can see. In the context of inflationary cosmology, it is fair to say that the universe is the ultimate free lunch."[74]

There are many theories which have no necessity for such a finite edge to the universe, and I will mention a few. I am not seeking to debunk the BGV (rather, to show its limits), though, as with every theory in modern cosmology, there are issues with integrating it with all other theories. However, it is worth noting that Sean Carroll, in his aforementioned debate with Craig, stated that there were over a dozen plausible models for the universe, and this included some eternal ones!

One such alternate theory is known as the Big Bounce Theory such that the Big Bang was simply the beginning of a period of expansion that followed a period of contraction. There are several manners[75] in which it has been suggested that this works. To me, the most intriguing theory is known as Loop Quantum Gravity[76] (LQG) which does not necessitate an initial singularity. The field of LQG[77] is very active and in its early stages, but it is defined by some of the most cutting edge physics taking place in the world. For example, Yongge Ma in his paper "The Cyclic Universe Driven by Loop Quantum Cosmology" in the *Journal of Cosmology* concludes, "It turns out that the classical big bang singularity will get replaced by a quantum bounce in all scenarios."

String theory (or variously superstring theory and M-theory) is another theory of gravity which attempts to be a theory of everything such that it marries the problematic pair of theories of general relativity and quantum mechanics. It claims that elementary particle such as electrons and quarks are not 0-dimensional but are 1-dimensional oscillating strings. There are also entities called branes (membranes), hence the name M-theory, and various numbers of extra-dimensions (M-theory requires 11 dimensions).

Then there is Steinhardt and Turok's "Endless Universe" theory, as well as Roger Penrose's "Cycles of Time". Indeed. Alireza Sepehri stated in his 2015 paper[78]:

> Recently, some authors removed the big-bang singularity and predicted an infinite age of our universe. In this paper, we show that the same result can be obtained in string theory and M-theory;

As reported earlier this year (2015)[79]:

The universe may have existed forever, according to a new model that applies quantum correction terms to complement Einstein's theory of general relativity. The model may also account for dark matter and dark energy, resolving multiple problems at once....

The physicists emphasize that their quantum correction terms are not applied ad hoc in an attempt to specifically eliminate the Big Bang singularity. Their work is based on ideas by the theoretical physicist David Bohm, who is also known for his contributions to the philosophy of physics. Starting in the 1950s, Bohm explored replacing classical geodesics (the shortest path between two points on a curved surface) with quantum trajectories.

In their paper, Ali and Das applied these Bohmian trajectories to an equation developed in the 1950s by physicist Amal Kumar Raychaudhuri at Presidency University in Kolkata, India. Raychaudhuri was also Das's teacher when he was an undergraduate student of that institution in the '90s.

Using the quantum-corrected Raychaudhuri equation, Ali and Das derived quantum-corrected Friedmann equations, which describe the expansion and evolution of universe (including the Big Bang) within the context of general relativity. Although it's not a true theory of quantum gravity, the model does contain elements from both quantum theory and general relativity. Ali and Das also expect their results to hold even if and when a full theory of quantum gravity is formulated....

In a related paper, Das and another collaborator, Rajat Bhaduri of McMaster University, Canada, have lent further credence to this model.

Craig is appealing to science, and cherry picking it to support his argument. One could accuse me of doing this. But I am not claiming these theories are right. Indeed, most of them must be wrong as they are probably mutually exclusive! However, I am not dogmatic about it. There are many plausible models out there, and many of them do not require a beginning to the universe the sort on which Craig depends for his argument, and thus for God. Indeed, as a non-physicist, I get the impression that the epistemic probability of the proposition, "The universe's expansion has/had a beginning," is very high (>90% or even >95%). But I do not get anything remotely like that when I consider the proposition, "The universe has/had a beginning." In fact, I can't even tell from reading various physicists if I think the probability of that latter proposition is greater than 50%! Whenever I read or hear a physicist talk about how the current laws of physics break down at the Big Bang (and how they are waiting for a new theory of quantum gravity to fix that), that makes me wonder if the current probability of "The universe has/had a beginning" is unknown.

Indeed, with regard to Craig's cherry picking, I get a sense of desperation, and this is perhaps borne out by his special pleading, cherry picking and science denialism, as can be seen in section 5.6.

Craig also invokes the second law of thermodynamics to defend the second premise. This, like the two a priori arguments regarding actual infinites (which are discussed by James East in his section, 4.1.2) looks at showing that the universe was not past eternal. I agree that this universe was not past eternal, but do not think this is relevant since at each Big Bounce or similar, time starts again. There is much to discuss in such cyclical or oscillating models. Essentially, the universe, as a brute fact, continues to exist, but each singularity or bounce infers a new creation event or similar such that there is not a linear regress of time into the past. Craig's use of the second law of thermodynamics looks to show, from an observably scientific viewpoint (rather than from an a priori mathematical one) that the universe cannot have an infinite regress into the past.

So what is this second law? The second law of thermodynamics states that in a closed system, following the arrow of time (something to which I will later return), the system will always move from a lower state of entropy to a higher state. In other words, the universe will get steadily colder, with energy dissipating, rather than being able to perpetually convert, say, energy to order. This is why perpetual motion machines are deemed impossible. As far as the universe is concerned, this implies, according to Craig, "given enough time, the universe will reach a state of thermodynamic equilibrium, known as the 'heat death' of the universe." (Craig 2002) Such an ending can either be hot, if the universe re-contracts, or cold if it expands forever. The implications of this, and the hows, whens and whys remain uncertain. There remain, as ever, alternative theories as to the ultimate fate of the universe. The basic idea is that if, eventually, the universe will hit a period of entropic equilibrium (heat death) and the universe regresses infinitely into the past, then why have we not reached this stage already? This can cause problems for certain cyclical models of the universe too, as Craig claims[80]:

> ...the thermodynamic properties of this model imply the very beginning of the universe which its proponents seek to avoid. For the thermodynamic properties of an oscillating model are such that the universe expands farther and farther with each successive cycle. Therefore, as one traces the expansions back in time, they grow smaller and smaller. As one scientific team explains, "The effect of entropy production will be to enlarge the cosmic scale, from cycle to cycle. . . . Thus, looking back in time, each cycle generated less entropy, had a smaller cycle time, and had a smaller cycle

> expansion factor than the cycle that followed it."
> Novikov and Zeldovich of the Institute of Applied
> Mathematics of the USSR Academy of Sciences
> therefore conclude, "The multicycle model has an
> infinite future, but only a finite past." As another writer
> points out, the oscillating model of the universe thus
> still requires an origin of the universe prior to the
> smallest cycle.

The issue with this view is that it is outdated. Cosmology is a rapidly changing discipline and the discovery of dark energy, for example, has fundamentally altered the playing field. The previously mentioned models, such as LQG, ekpyrotic (a cyclical or bouncing model) and Baun-Framptom models (another type of cyclic model), have featured prominently again, as a result. These models seek to answer problems with the standard Big Bang model (such as why the cosmological constant is several orders of magnitude smaller than predicted by the standard Big Bang model).

The simple truth of the matter is that I, no more than Craig, can claim that one model is vastly superior to another for explaining the ontology and history of the universe or multiverse. There are problems with *all* the theories, to greater and lesser degrees (otherwise everyone would agree on the model which had *no* problems!). However, it does seem that the cutting edge physics is taking place within the disciplines of some of these alternative theories (of, say, Loop Quantum Gravity and string theory) and one would be ill-advised to discount them so readily as Craig appears to.

4.5 The call for cosmological agnosticism

Scientists who look at the BGV conclusion and take it on board do not, as Craig does, take it as an answer, in and of itself, to one of the biggest cosmological conundrums, but see it as raising other questions. Singularities are points at which our knowledge of known science breaks down and there is much to still be grasped. The three scientists, Borde, Guth and Vilenkin, understand their theory far better than Craig does and are themselves non-theists. This should tell you something. There is enough we do not know, even if we take the BGV as given, to suggest a huge amount of elbow room for philosophical, theological and scientific unknowns.

Victor Stenger reported the Caltech cosmologist Sean Carroll as saying[81]:

> "I think my answer would be fairly concise: no result derived on the basis of classical spacetime can be used to derive anything truly fundamental, since classical general relativity isn't right. You need to quantize gravity. The BGV [Borde, Guth, Vilenkin] singularity theorem is certainly interesting and important, because it helps us understand where classical GR breaks down, but it doesn't help us decide what to do when it breaks down. Surely there's no need to throw up our hands and declare that this puzzle can't be resolved within a materialist framework. Invoking God to fill this particular gap is just as premature and unwarranted as all the other gaps."

A major issue faced with cosmology is the idea that someone reflecting on the various theories and fields *requires* a particular theory to be true in order to sustain a worldview or presupposed ideal. Admittedly, this can work both ways. A theist such as Craig necessitates something like the standard Big Bang model together with the Borde-Guth-Vilenkin theorem as well as other suppositions in order to allow for a conclusion that God created the universe. On the other hand, a non-theist might seek out theories which do not require a definite beginning in order to maintain their worldview.

Such desire to find evidence which supports a presupposed ideal is dangerous when appraising a discipline which is so changeable and in its nascent stages, as cosmology is. Dark energy was not empirically evidenced until 1998 and it accounts for a staggering 73% of the mass-energy of the universe. That "fact" alone is enough to warn anyone off being cosmologically dogmatic.

Cosmological dogmatism is something that William Lane Craig is certainly guilty of. In his necessity to have the BGV theorem as true, and its interpretation as implying a definite beginning to the universe (which, as we have seen, is arguable), Craig needs to put all of his cosmological eggs into one basket, and generously bet on it being the heaviest basket. This seems to be a result of his attachment to the Kalam Cosmological Argument and his desire for it to be sound, given the amount of time and effort that he has dedicated to it, and the number of debates and arguments that have, at least somewhat, hinged upon its veracity. There is not a little psychology involved with philosophy and science, I wager.

Cosmological agnosticism is something which is encouraged within the discipline of cosmology, as Timothy Eastman in *The Journal of Cosmology* states[82]:

> It is at present unknown how BB [Big Bang] or any of these alternative approaches will stand up to future tests using burgeoning new data sets and future critical tests and falsification instances. At the present time, I see both advantages and serious problems for all options – they may all be wrong – thus, the "agnosticism" in physical cosmology.

There are just too many theories vying for explanatory prominence within a field where new and ground-breaking data is coming in at a thick and fast rate. Philosopher Rem B. Edwards, on the other hand, proposes (in *What Caused the Big Bang?*) a stronger form of agnosticism, that we *cannot* know the answers of what caused the Big Bang[83]:

> Scientists now advance a variety of explanations for the origin and evolution of the universe; but agnosticism says that we really do not know the answers. General Cosmological Agnosticism does not deny that many cosmological puzzles can be resolved… What caused the Big Bang?…we do not know. We only know what happened after the Big Bang was inaugurated, and science cannot tell us what caused the Big Bang. Coming chapters will examine naturalistic theories that purport to offer scientific answers; but they actually give only highly speculative and dubious philosophical accounts of the ultimate origin of the universe, without acknowledging their subtle shift from science to unverifiable metaphysics.

The difference between these two positions may seem subtle, but it is epistemologically fundamental. Either we don't know because there are many competing theories, the discipline is nascent and huge areas of data and knowledge are *as yet* unknown; or we simply *cannot*, and therefore, *will never* know what caused the Big Bang. Both positions are agnostic, obviously, and there is some merit in both. However, I do not necessarily adhere to the notion that we *cannot* find the answer, or more definite an answer than what is proffered at the moment, I merely believe that we are in an early phase of knowledge in the field, and we need to be guarded with our inductive assertions, being open to alternative theories. The scientific model is the best epistemological model for moving from a lack of knowledge to a reliable bank of knowledge that we have. One could ask what religion or theology, as an epistemological system of sorts, over the last two thousand years has added to the bank of knowledge we have today. Compared to science, not a lot. Rem B. Edwards, quoted above, has a dig himself at William Lane Craig for his singular focus on the Standard Model[84]:

> The astute debate between William Lane Craig and Quentin Smith in their *Theism, Atheism, and Big Bang Cosmology* focuses almost exclusively on the Standard Big Bang Model, with its initial singularity, and on the quantum Big Accident option; but it neglects all other theories of origin explored here.

4.6 Naturalism as a good bet

Having established some good reason for remaining agnostic on the cause of the Big Bang, or what the exact processes were causally instrumental in the universe coming about or maintaining itself, I would also like to take this opportunity to present a case for having some self-defended intuition that the cause will be found to be naturalistic if found out at all.

The terms of supernaturalism and naturalism are incredibly hard to accurately pin down, but for the purposes of being concise with this point, let me assume "supernatural" entails a God causing the universe, and "naturalism" entails non-Godly machinations which adhere to natural laws. The basis for my approach in asserting that, from a probabilistic point of view, one would be unwise to bet against naturalism being the causal framework within which the universe "started" comes from Richard Carrier and Jeffery Jay Lowder. Philosopher Carrier set out in his 2006 written debate against apologist Tom Wanchick the following point[85]:

> The cause of lightning was once thought to be God's wrath, but [it] turned out to be the unintelligent outcome of mindless natural forces. We once thought an intelligent being must have arranged and maintained

the amazingly ordered motions of the solar system, but now we know it's all the inevitable outcome of mindless natural forces. Disease was once thought to be the mischief of supernatural demons, but now we know that tiny, unintelligent organisms are the cause, which reproduce and infect us according to mindless natural forces. In case after case, without exception, the trend has been to find that purely natural causes underlie any phenomena. Not once has the cause of anything turned out to really be God's wrath or intelligent meddling, or demonic mischief, or anything supernatural at all. The collective weight of these observations is enormous: supernaturalism has been tested at least a million times and has always lost; naturalism has been tested at least a million times and has always won. A horse that runs a million races and never loses is about to run yet another race with a horse that has lost every single one of the million races it has run. Which horse should we bet on? The answer is obvious.

The point was taken up and expanded by Jeffery Jay Lowder in what he calls the Evidential Argument from the History of Science (AHS). I have included this in the Appendix, which I appeal to you to read. The AHS takes the idea that naturalism has continually provided explanations which have superseded supernaturalistic ones, and that this explanatory supplanting has been one-way traffic only. As such, the prior probability for a supernaturalistic explanation to be false is high. In fact, we have pretty well begun, or succeeded, in explaining almost everything in the universe in terms of a naturalistic framework to the point that all we have left to explain, arguably, is consciousness and the start of the universe. Who would, indeed, bet against the naturalistic horse?

PART FIVE
The Syllogism's Conclusion

The conclusion, which follows necessarily (assuming none of the previous objections cause terminal problems to the argument before one gets to the conclusion) from the premises, is:

Therefore, the universe has a cause

Although it seems like a simple conclusion that would follow, without much ado, from the two premises, there are several problems with said conclusion. One could posit that the universe has caused itself in some manner, such as Daniel Dennett has done in *Breaking the Spell*. However, I will not spend any time looking at this scenario since it is not part of my objections to the conclusion. This cause, as Craig has extended the argument, is a personal God. Other evidence unconnected to the KCA is used to move the theist towards the Judeo-Christian version of the personal God. As Craig sets out with J.P. Moreland[86]:

1. The universe has a cause;
2. If the universe has a cause, then an uncaused, personal Creator of the universe exists, who sans the universe is beginningless, changeless, immaterial, timeless, spaceless and enormously powerful;

Therefore:

3. An uncaused, personal Creator of the universe exists, who sans the universe is beginningless, changeless, immaterial, timeless, spaceless and enormously powerful.

I will quote Craig at length here to spell out his own ideas involving how the conclusion should be interpreted[87]:

> It therefore follows that the universe has an external cause. Conceptual analysis enables us to recover a number of striking properties which must be possessed by such an ultra-mundane being. For as the cause of space and time, this entity must transcend space and time and therefore exist atemporally and non-spatially (at least without the universe). This transcendent cause must therefore be changeless and immaterial, since timelessness entails changelessness, and changelessness implies immateriality. Such a cause must be beginningless and uncaused, at least in the sense of lacking any antecedent causal conditions, since there cannot be an infinite regress of causes. Ockham's Razor (the principle which states that we should not multiply causes beyond necessity) will shave away further causes. This entity must be unimaginably powerful, since it created the universe without any material cause. Finally, and most remarkably, such a transcendent cause is plausibly to be taken to be personal.

I will now look to establish issues with such a conclusion.

5.1 Causality and time

If causality depends upon the notion of time, then if the universe with its spacetime did not exist, how could the process of causality be enacted in the creation of the universe?

Does, then, causality depend on time? David Hume seems to have thought so. In his *Treatise on Human Nature*, Part III, section XV, he set out that "The cause and effect must be contiguous in space and time." Causal lines are often defined in such a way that "the concept of a causal line can be used to explain the identity through time of an object or a person", as Phil Dowe explains of Bertrand Russell's reasoning on causality (Dowe 2007). It seems to me that time is a necessary paradigm within the confines of causation: time, in one sense, measures change; and something outside of time will have no change, since no causal force will be acting upon it.

If we don't have time, what do we have with regard to causation? Without time, we have simultaneous causation, a notorious problematic notion and one which proponents of the KCA need in order to satisfy the desire to make it a coherent argument. If two things occur simultaneously with each other, then one cannot define which one causes the other. William Wharton, a physicist at Wheaton College, concludes this, in his essay "Time and Causality":

> In closing, special relativity makes very clear that causation has a close connection with the time coordinate, unlike the spatial coordinates. There is no lower limit to the speed at which a causal chain can traverse space-time between two events. This means the two events can be at the same spatial location at different times. However the two causally connected events can not be at different spatial locations at the same time, because the maximum speed at which the causal chain traverses space time is the speed of light, c. Causal chains must traverse time but not necessarily space.

It does seem very problematic to take the behaviour of causation that we know only, and coherently, to operate in time, and then apply this to an a-temporal framework. As Jonathan Schaffer in his entry

91

"The Metaphysics of Causation" in the *Stanford Encyclopedia of Philosophy* declares about the notion of simultaneous causation:

> The main reply to the simultaneous causation argument is that the cases appearing to exemplify it are misdescribed (Mellor 1995). The iron ball takes time to depress the cushion, and in general all bodies take time to communicate their motions. There are no perfectly rigid bodies, at least in any nomologically possible world. Without the intuitive support of this sort of case, the simultaneous causation argument may be charged with begging the question.

The iron ball depressing a cushion is an analogy that Craig himself has used in public debates, and one that appears to be problematic. Grünbaum takes serious issue with Craig's defences of simultaneous causation, both symmetrical and asymmetrical, where asymmetrical causation explains how X can affect Y but not vice versa. Grünbaum emotively declares[88]:

> In any case, in the face of my demonstration above of the failure of Craig's argument for a simultaneous cause of the Big Bang, this Craigean declaration is unavailing. But he also tells us that there are mundane cases of instantaneous causation that qualify as asymmetric, one-way or directed. And he cites Kant's example of a ball denting a cushion. Another example is supposedly furnished by a locomotive pulling a train....

> It is a commonplace that, if the special theory of relativity is taken into account, a force applied to one end of a solid rod lying on a table will not affect the other end simultaneously, because the transmission of

an influence cannot be instantaneous in that theory. Similarly for Kant's cushion-example and the case of the locomotive. In a footnote, Craig dismisses this denial of instantaneous causation. I regret to say that I can discern only question-begging babble in what he says there.

Worse, for good measure, Craig adds that, according to some philosophers, "all efficient causation is simultaneous." It is the old maxim "Cessante causa, cessat effectus." The argument given for this claim as stated by Craig is a patent non-sequitur and yields a false conclusion, thereby illustrating the pitfalls of such a priori theorizing about causation.

The KCA certainly provided ammunition for these two philosophers to fire loud intellectual broadsides at each other. Let me investigate the arguments that Grünbaum uses and refers to in refuting simultaneous causation. It is interesting to note that I have not heard Craig use the ball and cushion argument or indeed any simultaneous causation argument in any of his more recent debates but he does espouse it in much of his writing. It is important to show the objections to simultaneous causation here because it is fairly crucial to showing the inadequacy of the conclusion that the universe has a cause. Without time existing "prior" to the Big Bang, the cause of the Big Bang must have been simultaneous with the Big Bang itself.

Firstly, and perhaps intuitively the strongest argument, is no matter the number of causes and effects claimed to be simultaneous, there still needs to be some explanation as to why the causal order of non-simultaneous causations is correlated to the time order. There is "no credible explanation of this difference that lets any causation be simultaneous" (Mellor 1998: 108). Special relativity also shows simultaneous causation to be false, as Grünbaum (1994) also shows, such that moving things transmit causation across space but they are restricted by the speed of light, thus inhibiting simultaneous causation. Another objection to simultaneous causation is upheld by Newton's third law of motion which states that the momentum of each of two colliding objects causes the other object's momentum to change. However, if this was simultaneous, this would make each object have changed and unchanged momentum *at the same time*.

For further arguments and development of these points and refutations of Craig's position on simultaneous causation, at least *within* the universe, see Grünbaum (1994) and, particularly, Mellor (1998: 108-111). Craig deals lightly (assertively but without much meat to the bone) in Craig (2009: 196).

5.3 Intentionality and time

Aside from the issue of simultaneous creation from a causal point of view, there is the connected problem that critics have for a long time held against a timeless God creating. God being able to intentionally create time seems to presuppose the existence of time (i.e. creation is a temporal process). Craig believes that "God is timeless without the universe and temporal with the universe"[89]. Craig attempts to answer the aforementioned problem by declaring[90]:

> On a relational theory of time, time is logically posterior to the occurrence of some event. So on a relational theory, God's acting is explanatorily prior to the existence of time. All God has to do is act and time is generated as a consequence. So God could both create *t* and exist at *t*.

Craig continues his proposition by stating that[91]:

The claim that if God is timeless, it is impossible for Him to create the universe is based upon the assumption that timelessness is an essential, rather than contingent, property of God. But as in the case of the color of the house, I see no reason to think that God's being timeless or temporal cannot be a contingent property of God, dependent upon His will. Existing timelessly alone without the universe, He can will to refrain from creation and so remain timeless; or He can will to create the universe and become temporal at the first exercise of His causal power. It's up to Him.

I have a problem with this idea since willing to do something (entering time by creating) is itself a temporal process, and so I think Craig's theory falls down here. This is moving towards the area of debating divine personhood and what, ontologically speaking, makes this up, and whether it is possible within a timeless context. Craig has worked on this at length, though I remain unconvinced by his case. For more on his views, I would direct you towards his essay "Divine Timelessness and Personhood" and his book *God, Time, and Eternity: The Coherence of Theism II: Eternity*. The primary issue is that for God to have the intention to create and creating, as Craig believes, this surely requires time. There is a change here and that requires time, being (in a sense) that time itself is a measurement of change. It is not as if God creates time at creation, because prior to this comes intention. Intending, as far as I am concerned, requires time. This can be summed up as a logical syllogism:

1) God is the only object which existed (causally or explanatorily) prior to the creation of the universe
2) God is ontologically perfect
3) Perfection entails no needs or desires, and is a state which cannot be improved

95

4)   God could not have intended the universe to exist prior
to the creation of the universe

So the problems here for the God thesis are similar to the issues
with causality and are twofold: one of intentionality and one of
perfection. To take the first issue, what we have is the idea that God
having intention to create the universe causally prior to time is
incoherent. We have looked at time and causality in the last section.
I will focus on intention in this section. Craig's defence of God
creating spacetime intentionally can be summed up as follows. He
believes that exactly the same reasoning that he applies to causality
can be applied to intentionality: that of simultaneity. Craig used the
following analogy, with desperate similarities to his other analogies
used in the last section:

> There's no reason that the intentions, and the actions
> that are the result of those intentions, couldn't be
> simultaneous. For example, think of someone who was
> dangling off of a cliff, hanging onto a tree root, to avoid
> falling. He intends to hold onto that root as firmly as he
> can and his intentions are not *chronologically prior* to
> the action of fastening his hands around the root:
> they're simultaneous. So there simply isn't any need to
> have a chronological priority of intention to action. I
> think what is prior is a kind of *explanatory priority*: he
> holds onto the root tightly because he doesn't want to
> fall. So there is an explanatory priority to intention to
> action, but it doesn't need to be a chronological
> priority.[92]

There are some critical objections that can be made to such a
defence that intentionality does not necessitate time and can be
simultaneous.

Firstly, Craig is conflating two ideas of intention into this analogy, as discussed variously by Justin Schieber and Dr. John Danaher[93]:

(IC) Intentions that change a state of affairs
(IM) Intentions that maintain a state of affairs[94]

In the cliff-hanger analogy, he espouses simultaneous intentionality and action as allowing for IM, but implicitly applies it to examples of IC. In other words, he uses IM to define *all* intentionality. Whilst intentionality can be simultaneous with action to maintain a state of affairs, I contest that it can be simultaneous with action in order to change a state of affairs. This would entail there existing, all in one "instance", the intention to create the universe, the state of affairs without the universe *and* the state of affairs *with* the universe (since in order to intend there to be a universe, that universe must not exist). The change inherent in moving from nothing to something, from creating *ex nihilo*, necessitates time, as well as moving from intention to actualisation. The only way IM can be seen to be effective in the state of affairs of the universe is if the universe had always, eternally (in an infinite amount of time sense) existed, which is precisely *not* what Craig argues!

Secondly, the analogy still fails to overcome the causal objections from the previous section. The man intends to keep his fingers tightened. This is a continual action, but it *does* have a start. At the start, a signal is sent from the brain to the fingers and muscles. Signals travel, muscles expand and contract. These signals are continually sent. There *is* temporal priority here.

As a result, the idea that God could intend to create the universe (spacetime) a-temporally, and create it *ex nihilo* is incoherent from both a causal perspective and from an intentional one.

As far as perfection is concerned, there are problems. Suffice to say that I adhere to the school of thought which declares that ontological perfection, as God held "prior" to creating the universe, would not entail the desire or intention to create the universe (irrespective of temporal, causal or intentional issues), especially since God still has ontological perfection "after" creation and throughout the temporal process of the world, with all its supposed issues. If God was and still is perfect, what need, or why intend the creation of the world? And so, we move sideways on to the next section, which looks in more detail at this.

5.4 The Argument of Non-God Objects

I am going to go a little off-piste here in challenging the notion that God would create anything at all, as hinted at in the section in intentionality. This looks to undermine the KCA from a position of a god as understood within the realms of classical theism, that is, a god who is omniscient, -potent and -benevolent.

Let us return to nothingness and God. If there was nothing but God, then what good reason could God have for creating us, that thing there, cancer, some fluff, or, well, anything? When you see God as an ontologically perfect state of affairs, then it appears that everything else is a degradation of that. (Unless I am also the exemplification of ontological perfection. I like to think so. No one else seems to agree).

This leaves us with an uncomfortable conclusion: that if God was perfect, then he/she/it would likely not have created anything at all, least of all this here universe.

The arguments seeks to show that there is a logical incompatibility between the notion of a classically theistic God (in its maximal greatness), and the existence of anything, namely the world. Such a perfect God is defined within the discipline of what is called Perfect Being Theology (PBT).

PBT states that if God exists, then he has the maximally compossible degrees of greatness, such as power, knowledge and benevolence.

Christian philosopher, J.P. Moreland states[95],

> To say that God is perfect means that there is no possible world where he has his attributes to a greater degree... God is not the most loving being that happens to exist, he is the most loving being that could possibly exist so that God's possessing the attribute of being loving is to a degree such that it is impossible for him to have it to a greater degree.

The argument simply states that if God is the greatest conceivable entity to which nothing could compare, then he would have no need or desire or necessity or reason to create anything non-God.

Indeed, Benedict de Spinoza argued something similar in the appendix to Part 1 of his Ethics:

> Further, this doctrine does away with the perfection of God: for, if God acts for an object, he necessarily desires something which he lacks. Certainly, theologians and metaphysicians draw a distinction between the object of want and the object of assimilation; still they confess that God made all things for the sake of himself, not for the sake of creation. They are unable to point to anything prior to creation, except God himself, as an object for which God should act, and are therefore driven to admit (as they clearly must), that God lacked those things for whose attainment he created means, and further that he desired them.

Justin Schieber, a skeptical philosopher and debater, defended this argument in debate with Canadian apologist Michael Horner.[96] This is what such an argument would look like as set out in a logical syllogism[97]:

> P1: If the Christian God exists, then GodWorld is the unique best possible world.
>
> P2: If GodWorld is the unique best possible world, then the Christian God would maintain GodWorld.
>
> P3: GodWorld is false because the Universe (or any non-God object) exists.
>
> Conclusion: Therefore, the Christian God, as so defined, does not exist.
>
> *Note: The term 'GodWorld' refers to that possible world where God never actually creates anything. That God's initial act of creating the universe (or any non-God object) was an act not borne of necessity is an implicit assumption within this argument.*

Not only are there problems with the temporal notion of desiring or intending and then acting on those intentions, and moving from nothing to something, in an a-temporal paradigm, but there is the notion that this would also be a degradation of a perfect state of affairs. A perfect god, it would seem, would not want to degrade perfection and create that which is less than perfection. This universe. Us.

5.5 Simultaneity as a temporal notion

Getting back to simultaneity, a further objection to both the causal and intentional use of simultaneity by Craig is the idea that simultaneity is itself a temporal term and idea. All definitions seem to be contingent upon the idea of time (such as simultaneous meaning "occurring, existing, or operating at the same time" as accrding to the World English Dictionary[98]). The problem is exacerbated by the Theory of Relativity which implies that it is impossible to say whether two distinct and separate events take place at the same time if they are themselves separated by space (in any kind of absolute sense). Time is a relational notion, and the sense that there is an absolute time at which two things can happen is incoherent other than under a commonsense usage and understanding.

It is difficult to get one's head round the terminology and conceptuality of something like time, but in order for Craig to have a robust defence of the KCA, he must be able to coherently establish how God can have both simultaneous causality and intentionality with the creation *ex nihilo* of the universe. As it stands, he seems to be building his argument on shaky ground.

A friend of mine, known as Counter Apologist (isn't it a shame, and a reflection of the cultural bias against atheism, that people still have to live under pseudonyms in supposedly progressive countries), has put together a strong argument against Craig's use of time and science, as follows in the next section. What Craig insists upon, is using an understanding of science, an interpretation of time that scientists do not themselves adhere to in order for his version of the KCA to work. This is pertinent because he insists on using science and scientific theories, such as the BGV, to defend his use of the KCA in other ways. In other words, he is seriously cherry picking his science and appears to be a strong advocate for double standards. This alone should invalidate his use of the KCA.

This next section does include some technical language and points. It is a large and complex area, time and the philosophy and science concerning it, and, as such, all points cannot be fully explained here. Thus it is suggested the reader make further investigations into the subject to support better understand the section, if necessary. The section is longer than other sections, though, in order to present a robust case for the weakness of Craig's approach.

I think Counter Apologist's points are extremely important and are fairly terminal for Craig's case.

5.6 On time: Craig's inconsistent appeals to science, by Counter Apologist

Unlike most of the arguments for the existence of a god, the Kalam's main appeal is that it purports to show that modern science points to the existence of a god.

This is a bit of a shock to most cosmologists and physicists, since if you were ever to attend one of the many cosmology conferences around the world, you'll hardly find any reference to a god as an explanation. In fact, two of the scientists Dr. Craig likes to reference in his presentation of the Kalam—Alexander Vilenkin and Alan Guth—expressly do not believe in any kind of a personal creator god.

It's a bit curious when we see a theologian start talking about modern cosmology as evidence for the existence of a god, when some of our most prominent physicists who produced that knowledge are themselves atheists.

This isn't proof that the Kalam is false, but it is something that should cause us to be skeptical of the argument. My goal in this section is to show exactly why this disparity exists between the physicists who study cosmology and the theologians who co-opt it for use in modern apologetics.

One of the key differences between physicists and theologians getting wildly different conclusions from the same set of data boils down to how each group thinks about time.

In contemporary philosophy, there are three positions on how we think about time:

- The A-Theory of Time—A "tensed" theory of time. A tensed statement would be something like "It is cold today", because it depends on the temporal perspective of the person who says it.[99]
- The B-Theory of Time—A "tenseless" theory of time. A tenseless statement would be something like "It is cold on December 16 2015".[100]
- Time is not real or fundamental—The idea that time itself isn't a fundamental component of reality, but emerges from a more basic set of laws.

A detailed discussion of the different views of time is well beyond the scope of this book, however for readers looking for more detailed information would be served by checking out the fantastic entry on time in the Stanford Encyclopedia of Philosophy.[101]

For our purposes here, it is enough to say that the Kalam Cosmological Argument is entirely predicated on the A-Theory of time. This is admitted as much by William Lane Craig[102]:

> From start to finish, the kalam cosmological argument is predicated upon the A-Theory of time. On a B-Theory of time, the universe does not in fact come into being or become actual at the Big Bang; it just exists tenselessly as a four-dimensional space-time block that is finitely extended in the earlier than direction. If time is tenseless, then the universe never really comes into being, and, therefore, the quest for a cause of its coming into being is misconceived.

As Dr. Craig alludes to, if we reject the A-Theory and accept the B-Theory of time then the Kalam loses its force. Elsewhere in his work, Dr. Craig admits that the A-Theory is rejected by a majority of physicists. This explains why we see so much of a difference between the two camps in their interpretation of modern cosmology.

The interesting question that follows from this is to ask why do physicists reject the A-Theory, and why do theologians like Craig accept it?

The main problem for the A-Theory of time is that it is largely considered to be incompatible with Einstein's Special and General Relativity (STR and GTR). In order to work around this issue, Dr. Craig advocates for what is known as a "Neo-Lorentzian interpretation" of Special Relativity.

My contention here is that if we follow the methods of science, it tells us that the Neo-Lorentzian interpretation, and hence the A-Theory, is most likely false.

Dr. Craig thinks he can work through this problem by rejecting the conclusions of what science tells us by instead relying on purely philosophical, metaphysical arguments for the A-Theory. For what it's worth, I think those metaphysical arguments fail, which I will cover later in the section, but for now my point is to expose a key flaw in Dr. Craig's appeal to science in the Kalam.

Dr. Craig is as free as he wishes to reject the conclusions of science, and to rely on purely philosophical arguments for whatever conclusion he wants to establish.

What he is not free to do however, is to reject the conclusions of science in one context, but then appeal to them in another context when it is convenient. The best you could say about that is that it is cherry picking. We can see the evidence of this here in his debate with Sean Carroll[103]:

> "The evidence of contemporary cosmology actually renders God's existence considerably more probable than it would have been without it. [...] I'm saying that

contemporary cosmology provides significant evidence in support of premises in philosophical arguments for conclusions having theological significance. For example, the key premise in the ancient Kalam Cosmological Argument that the universe began to exist is a religiously neutral statement which can be found in virtually any textbook on astronomy and astrophysics. It is obviously susceptible to scientific confirmation or disconfirmation on the basis of the evidence. So to repeat, one is not employing the evidence of contemporary cosmology to prove the proposition that God exists, but to support theologically neutral premises in philosophical arguments for conclusions that have theistic significance."

To paraphrase Craig, he is using the evidence of contemporary cosmology to support theologically neutral premises in philosophical arguments that have theistic significance.

The problem for him is that if we follow the evidence of contemporary cosmology (which follows the methods of science), the A-Theory of time is most likely false, and so his key premise in the Kalam argument is undermined.

To put this into a syllogism:

1. The evidence of contemporary cosmology renders the A-Theory of time most likely false
2. The evidence of contemporary cosmology is true (an assumption Craig makes).
3. Therefore, A-Theory is most likely false.
4. "The universe began to exist" is true if and only if A-Theory is true.
5. Therefore, "the universe began to exist" is most likely false.

Before getting to the meaty first premise, allow me to quickly explain why the fourth premise is true.

To understand premise four, we need to look to Dr. Craig's own definition for the phrase "beings to exist"[104]:

An entity *e* comes into being at time *t* if and only if

(i)   *e* exists at *t*,

(ii)   *t* is the first time at which *e* exists,

(iii) There is no state of affairs in the actual world in which *e* exists timelessly

(iv) *e*'s existing at *t* is a tensed fact

It's that fourth part that's the key here, since "tensed facts" only exist on the A-Theory of time. In fact that entire part is there to explicitly draw out that the Kalam is predicated on the A-Theory of time.

So according to the methods of science, the philosophical premise "the universe began to exist" is simply false, even if the universe had a beginning. This is because even in that case, the "beginning" is like the front edge of a ruler. The ruler is always there, even if there is a "first inch" marked on it. Basically, time doesn't work in the way Craig needs it to in order to argue for the existence of a god.

It is very important to note that this does not mean that science proves god does not exist. It simply shows that the impetus Craig is trying to use to argue for the existence of a god is false.

But does science really show that the A-Theory of time is most likely false?

At this point, critics might acknowledge my fourth premise, but take issue with the first.

For reference, you can substitute "the evidence of contemporary cosmology" for "science" in the argument; it works either way. This is because the evidence of contemporary cosmology is predicated on

Einstein's relativity, which is the very theory that shows us that the A-Theory is most likely false.

Since I intend this to also be a bit of a science lesson, let me define my terms a little: a privileged reference frame in terms of relativity in physics is essentially a physical place where the laws of physics work differently. Back in Einstein's day in the early 1900's this was known as the Aether, today you'd hear it called something of a place of "absolute rest" or the "absolute reference frame".

Now what Craig and other A-Theorists would have you believe is that when it comes to Relativity, and Special Relativity in particular, it is "simply a matter of taste" when it comes to whether you use the orthodox interpretation where there is no privileged reference frame or the Neo-Lorentzian interpretation where there is an undetectable privileged reference frame.

The truth of the matter is that science is not silent on this, and for good reasons.

Contrary to what Dr. Craig alleges, this isn't simply because all physicists and philosophers of science are holding to some outmoded form of logical positivism or verificationism (simply put, the idea that only things verified by empirically observable features are meaningful; the reader is welcome to look further into this elsewhere). Dr. Craig likes to make a lot of noise about the fact that the reason the Neo-Lorentzian view was discarded back in Einstein's time was for this simple reason. This is because, at the time, Einstein and other scientists actually did hold to verificationism or positivism, which has shown itself to be untenable in modern times. When criticism comes up he invariably brings this card out and accuses his critics of unwittingly being a positivist. What he doesn't talk much about is why in light of the failure of positivism modern science still holds to the standard interpretation of relativity and still disregards the Neo-Lorentzian view that Craig needs in order for the Kalam to work.

This is where science and philosophy of science come together to provide us with ways to decide between the sorts of interpretations on a set of empirical data that is at issue when it comes to relativity in this context. In modern science it is not enough for a theory to simply be consistent with the data. We can come up with myriad theories to simply "fit the data" to get whatever conclusion we want—including that the moon is made of green cheese.

Before getting into why modern science rejects the Neo-Lorentzian interpretation, let's do a very quick overview of what the evidence for relativity is, and then contrast what Neo-Lorentzian view entails compared to the standard interpretation of relativity.

The Evidence

I've spoken elsewhere[105] about some strong observational evidence we have for relativity: Time Dilation and Length Contraction. In short, the standard interpretation says that as we approach the speed of light, clocks of all types slow down uniformly and measuring rods contract in length. This sounds pretty crazy at first, but the fact is that we have a vast amount of experimental evidence for this. Our modern GPS systems are based on this very principle. The difference between the two views discussed comes down to what it means for "time to slow down" or for "measuring rods to contract".

The Standard/Einstein Interpretation

Einstein's relativity is based on two assumptions:
1. The speed of light in a vacuum is constant.
2. The laws of physics are the same in all reference frames (i.e., everywhere in the universe)

That second assumption is the key point of contention; it is often referred to as "Lorentz Invariance" or "the principle of relativity". In technical terms, it means that for any experiment we conduct, the results will be the same, regardless of: how we are oriented (rotation); translation between reference frames (i.e., different points of view observing the experiment); or how fast we are moving.

The Neo-Lorentzian Interpretation

Eventually, the "Neo-Loretnzian" interpretation has been derived down to two assumptions:

1.   There is a (undetectable) privileged reference frame with respect to which the speed of light in a vacuum is constant in all directions.

2.   The rates of electromagnetic clocks moving with constant speed v relative to the privileged reference frame all vary with v in the same manner.

I say "eventually" since the original Lorentzian approach to relativity was considered *ad hoc* since Lorentz first postulated that the ticking of "electromagnetic clocks" varied with velocity relative to the privileged frame. Then the theory had added to it the assumption that mass varied with velocity relative to the privileged frame to account for gravitational clocks slowing in the exact same way. And then the same modification with the weak nuclear force was necessary to account for meson decay experiments, and so on. It was finally in the 1950s that H.E. Ives was able to use the laws of conservation of energy along with these assumptions to derive an observationally equivalent set of equations to Einstein's.

For reference, the assumptions I'm quoting here are from S.J. Prokhovnik's derivation of the equations since I've been unable to locate a copy of Ives's derivation; however, Craig claims in his published work that Ives was able to use only two assumptions along these lines and so I take him at his word on this.

Accuracy

The first criteria we can use to decide between these two interpretations is accuracy. Clearly, science is justified in favoring a particular theory if that theory is more accurate than its competitors. Unfortunately for the context of our debate, both interpretations are on equal footing here.

Thanks to the work of Ives, for any experiment we conduct, the observational results will be the same regardless of which interpretation we hold.

The difference is that the Neo-Lorentzian view assumes that instead of the laws of physics being the same in all reference frames, there is a special "privileged" reference frame where physics behaves differently. In both interpretations the equations describing what we actually observe work out exactly the same way, and those equations end up being "Lorentz Invariant" for everywhere in the universe except for the supposed privileged reference frame assumed by the Neo-Lorentzian view.

This privileged frame is undetectable because it is our changing velocity relative to this privileged frame that causes us to observe the phenomenon of time dilation and length contraction.

Leading to New Advances

This is a very problematic area for the Neo-Lorentzian view. It is widely regarded that the key advance of Einstein's relativity was stipulating that the laws of nature must be fundamentally Lorentz Invariant. By assuming the laws of nature are Lorentz Invariant, science was able to make tremendous advances in seemingly unrelated areas. First was the advance from Special Relativity to General Relativity, then, by specifying Lorentz Invariance as a precondition, we were able to make huge advances in both quantum mechanics and quantum electrodynamics.

Contrast this with the Neo-Lorentzian view where the equations are not fundamentally Lorentz Invariant, they are only Lorentz-Invariant in terms of parts of the universe that we can observe. On this view, it would not have given physicists the same kind of clues to specify Lorentz Invariance as a precondition for all other fundamental physical theories which led to the advances mentioned above.

This is a major strike against the Neo-Lorentzian view.

The only objection I can think of here is if Craig were to assert that while the laws of physics were not fundamentally Lorentz Invariant, it would be somehow fundamental that they would always "appear to be". This would be an *ad hoc* modification of the Lorentzian view to avoid the problems laid out above, and it would also make the next problem all the more acute for the theory.

Simplicity

This is the nail in the coffin for the Neo-Lorentzian interpretation, and it is divided into two parts.

a) Unnecessary Entities

One does not have to be logical positivist or a verificationist to also hold that scientific theories which postulate extra entities which are unnecessary to explain all empirical data are more likely to be false than the simpler alternatives that do the same job.

b) Conceptual Simplicity

Proponents of the theory are quick to point out that, thanks to the work of H.E. Ives, the Neo-Lorentzian interpretation was able to reduce down to the same number of assumptions as Einstein's interpretation. However, the measure of simplicity isn't only about the number of assumptions required, although that is part of it. A more important measure is that of conceptual simplicity, which is where Einstein's interpretation is the clear winner.

Let us consider the assumptions. On Einstein's view we have a universe where the laws of physics are the same everywhere in the universe, and the speed of light in a vacuum is constant (or more specifically, no information can be transmitted faster than the speed of light).

When we move from theory to empirical investigation of the universe, this is exactly what we find to be the case. We would expect to observe the laws of nature to be Lorentz Invariant if we assumed the laws of physics were the same everywhere.

Contrast this with the Neo-Lorentzian view where fundamentally, the laws of nature are not the same everywhere in the universe, and that there is an absolute state of rest and progression of time. On this view, we would expect to find this in experiments, but we find the opposite. It is only by assuming that there is only one undetectable physical place in the universe where the laws of physics work differently than in every other place in the universe that the Neo-Lorentzian view is able to maintain compatibility with observation.

A great analogy is to "The Dragon in my Garage" as described by Carl Sagan in "The Demon Haunted World"[106]:

> A fire-breathing dragon lives in my garage" Suppose…
> I seriously make such an assertion to you. Surely you'd want to check it out, see for yourself. There have been innumerable stories of dragons over the centuries, but no real evidence. What an opportunity!
> "Show me," you say. I lead you to my garage. You look inside and see a ladder, empty paint cans, an old tricycle—but no dragon.
> "Where's the dragon?" you ask.
> "Oh, she's right here," I reply, waving vaguely. "I neglected to mention that she's an invisible dragon."
> You propose spreading flour on the floor of the garage to capture the dragon's footprints.
> "Good idea," I say, "but this dragon floats in the air."
> Then you'll use an infrared sensor to detect the invisible fire.

"Good idea, but the invisible fire is also heatless."

You'll spray-paint the dragon and make her visible.

"Good idea, but she's an incorporeal dragon and the paint won't stick." And so on. I counter every physical test you propose with a special explanation of why it won't work.

Now, what's the difference between an invisible, incorporeal, floating dragon who spits heatless fire and no dragon at all? If there's no way to disprove my contention, no conceivable experiment that would count against it, what does it mean to say that my dragon exists? Your inability to invalidate my hypothesis is not at all the same thing as proving it true. Claims that cannot be tested, assertions immune to disproof are veridically worthless, whatever value they may have in inspiring us or in exciting our sense of wonder. What I'm asking you to do comes down to believing, in the absence of evidence, on my say-so.

When you look into the garage, you see exactly what you would expect if there were no dragon. So it is with our universe, we see exactly what we'd expect assuming there is no place where the laws of physics worked differently than anywhere else.

Theism makes this problem worse

As a hypothetical, let's assume that theism is true and that God wants to leave breadcrumbs to lead people to his existence via natural theology. Presumably, if the Kalam was valid, God would be rather happy with apologists and the argument would fit very well with modern scientific investigation into the nature of physical reality.

But that's not what we see! What we see is that, by all accounts, our Space-Time universe at least appears to be Lorentz-Invariant. This entire debate over the theory of time simply wouldn't exist if

113

science actually revealed a world as scientists in Newton's time thought it was—one of absolute space and time. Experiments would simply be Galilean Invariant and we would never observe time dilation or length contraction, or we would simply be able to observe the effects of the ether/preferred reference frame.

Indeed, one must wonder why Craig's god would create a universe with absolute space and time, but do everything it can in order to make space-time's true nature appear to be the opposite!

The Scientific Verdict on Relativity

The philosophical criteria we use to determine between two scientific theories clearly favors the Einsteinian approach to relativity over the Neo-Lorentzian approach. I want to stress that this is in the absence of logical positivism or verificationism.

Without the Neo-Lorentzian approach, the A-Theory of time that the Kalam needs is shown to be scientifically untenable.

The final word?

The approach science and philosophy of science give us still allows for a way for the Neo-Lorentzian approach to win out "in principle" over the standard interpretation. All we would need is a set of empirical data that could only be uniquely explained if there were an undetectable privileged frame. There is ample opportunity for this to occur in finding a solution to the quantum gravity problem facing physics today. If doing so entailed a violation of Lorentz Invariance, then everything I've said would be overturned.

Dark matter is a great example of something in contemporary science that we would say is currently undetectable that we have good scientific reasons to think exists. When it comes to relativity, science is currently investigating whether or not there are violations of Lorentz Invariance in the laws of physics or in experiments and so far has found none.

This provides no succor to Craig, since science's provisional nature can just as easily invalidate any supposed evidence he appeals to in order to pretend that science points to the "beginning" of the material universe.

Science makes it worse for the Kalam

In his debate with William Lane Craig, Sean Carroll gives us yet another scientific reason to consider that science undermines the idea that the "universe began to exist". The piece of evidence is called the "Quantum Eternity Theorem" which states that if the total amount of energy in the universe is greater than zero, the universe must be eternal into the past and the future. It also states that if the total amount of energy in the universe is actually zero, then time itself is not a fundamental part of reality, and so the Kalam argument fails. Either way, the Kalam is in trouble.

Craig never responded to this point in the formal debate, but here is what he had to say on the second day of talks when pressed by Sean Carroll[107]:

> "I would say that time is one of the most evident realities to us, inescapably real. The reality of time is even more evident than the reality of the external world. Because I could be a brain in a vat, with illusions of an external world around me, but the stream of contents of consciousness in succession one after the other is undeniable and inescapable. Even the illusion of temporal passage is temporal passage. So that the reality of time, it seems to me, is one of the most basic, undeniable realities of metaphysics, of ontology, that there could possibly be. And if time does not appear on the fundamental quantum level, then so much the worse for the ontology of that level. Then that simply means

that it doesn't capture reality fully to speak of reality on
that sort of a scale."

Notice the response: Craig doesn't deny the theorem. He simply
says that in the only scenario where quantum mechanics allows for
the universe to be finite, then so much the worse for science's ability
to describe reality! He rejects the scientific conclusion that time is
ether eternal or not fundamental depending on what the total energy
in the universe actually is. Note that this doesn't entail that time is
not necessarily fundamental, it simply means time could not be as
Craig needs it.

What is his possible justification for such a dramatic claim? He
says that even if he was a brain in a vat, he would still be experiencing
his consciousness as a stream of temporal events, and so time must
be fundamental.

The problem with this of course is that for that to work, Craig
must assume that mind is fundamental. His metaphysics allows for
no method for consciousness to even possibly be emergent, and so
likewise time could not possibly be emergent.

This blatantly begs the question against naturalism, which would
assert that whatever the ultimate nature of reality is, it is material.
Contrast this with supernaturalism, which would assert that the
fundamental nature of reality is mental.

Scientific Conclusion

I believe the arguments above establish conclusively that when
it comes to appeals to current science with respect to the Kalam, it
tells us that the Kalam is more likely to be false than true. Proponents
of the Kalam cannot have it both ways. They cannot point to general
relativity and the standard big-bang model to say the universe had a
beginning but at the same time ignore what the methods of science
tell us about the nature of time in light of relativity theory. What's
more, as has been shown earlier in this book, that even pointing to

the standard big-bang model wouldn't necessarily establish that the universe began to exist, even if we assumed an A-Theory of time.

Perhaps a proponent of the Kalam is willing to admit this much but would then insist that metaphysical arguments can overturn the conclusion we arrive at via purely scientific criteria. I've already touched on how this is problematic in one way with the Quantum Eternity Theorem, but there are even bigger metaphysical problems for William Lane Craig.

Problems with Metaphysical Arguments for the A-Theory

William Lane Craig insists that metaphysical arguments can trump scientific conclusions. He will generally reference what he calls the two strongest arguments for the A-Theory of time:

1. The indispensability of tense from human language and thought
2. The incorrigible experience of the presentness of our own experiences

His first argument is true, we can't dispense tense from human language; however, this in itself doesn't do much for his case. This is because B-Theorists such as D.H. Mellor have shown that "although tense cannot be eliminated from our language, the truth conditions of tensed sentences need only tenseless facts, thus blocking need of an appeal to tensed features of reality."[108]

As for his second argument, I find this to be particularly poor. Consider the "incorrigible experience of our lack of motion while standing still," This doesn't detract from the fact that even while we are "standing still," we are still moving at incredible speeds through space-time as the earth moves around the sun, let alone the motion of our galaxy through space-time.

The list of "incorrigible experiences" that science has disabused us of is incredibly long. Just because we feel a subjective "now" doesn't mean that our consciousness, whatever it is, could not possibly be moving along some path in a 4D space-time.

However, I question our "experience" of the present moment on philosophical and scientific grounds. What does it mean to say we experience the present moment if any notion of an "absolute present" is in an undetectable privileged reference frame? Further, how long does the "present moment" last?

We very quickly run into problems trying to define that; in fact, the only principled answer I can think of would be a Planck second, which is 10^{-43} seconds. But this is well beyond the range of what we meaningfully "experience" in terms of the passage of time.

In fact, science tells us that we each live about 60—80ms in the past. Moreover, we will identify events that occur within that time frame around us as happening "simultaneously." We can't even finish saying the word "now" before it is no longer technically that time, so references to "now" are to an unspecified length of time relative to when the word was said or thought.

Let me be clear, one need not necessarily embrace the B Theory if you reject the A-Theory of time. All that we require is that the A-Theory of time to be false for the Kalam argument to fail.

Pure Metaphysics gives no Answer

Like most of the perennial questions in metaphysics, the question of time has boiled down to competing intuitions. None of the varied theories of time are incoherent, and based on these responses to the "strongest metaphysical arguments" for the A-Theory of time, I believe it is far from clear that we should prefer the A-Theory from a metaphysical perspective. This is borne out in our examination of views on the theory of time across disciplines.

When it comes to physics, one physicist has told me they know of no working physicist who holds to Craig's Neo-Loretnzian interpretation. Even Craig admits in his published work that the vast majority of scientists do not adhere to the A-Theory of time, but what about philosophers?

One of the most comprehensive surveys of philosophers was the 2009 Philpapers survey[109], here is what they found on the Philosophy of time:

> Time: A-Theory or B-Theory
> Other   542 / 931 (58.2%)
> Accept or lean toward: B-theory   245 / 931 (26.3%)
> Accept or lean toward: A-theory   144 / 931 (15.5%)

As you can see, A-Theorists are in the clear minority. Admittedly B-Theorists don't fare much better since the clear majority accepts another stance on the theory of time. However, this is still very problematic for Craig and the Kalam since unless you accept the A-Theory and the absolute present moment, the Kalam fails. For the record, I also am not explicitly arguing for the B-Theory per se; however, it does fit our current scientific picture more than any other view at the moment.

Remember, this is among professional philosophers. Still, we can glean more information from looking at the philosophers in more detail by sorting by Area of Specialty:

For Philosophers of Science we see the B-Theory pull ahead significantly and the A-Theory falls a bit:

> Accept or lean toward: B-theory   30 / 61 (49.2%)
> Other   24 / 61 (39.3%)
> Accept or lean toward: A-theory   7 / 61 (11.5%)

The inverse happens when we sort by Philosophers of Religion, a field other studies show to be overwhelmingly populated by Christians:

> Accept or lean toward: A-theory   19 / 47 (40.4%)

119

Other	18 / 47 (38.3%)
Accept or lean toward: B-theory	10 / 47 (21.3%)

I think this says quite a bit, especially considering in his response to me Craig says that his position on the theory of time is independent of his theological positions. I find that in reading arguments for the A-Theory I almost usually find it to be theists defending the position, with very few exceptions.

Let's look at one last area, Philosophers who specialize in Metaphysics:

Accept or lean toward: B-theory	98 / 234 (41.9%)
Other	80 / 234 (34.2%)
Accept or lean toward: A-theory	56 / 234 (23.9%)

This last fact is particularly interesting given Craig's defense of placing metaphysical assumptions over what science reveals to us about reality. Even among metaphysicians the B-Theory wins out.

Let me be clear yet again: absolutely none of this shows that B-Theory is true, or even that A-Theory is false. What it does show is that at best the argument within the philosophy of time is very far from settled and the situation isn't as clear cut as Craig likes to imply in his published works. The lack of physicists who hold to the A-Theory would also explain why so very many of them are atheists and why there is no talk of god at conferences on contemporary cosmology. You don't find the words "transcendent cause" in a cosmology textbook; what you find are *differential equations*!

What I intend to do next is to show the metaphysical problems with the A-Theory of time when it comes to the Kalam argument.

What is Time?

The Kalam is an argument that states that "time itself" must have had a beginning; but what exactly is "time" when we are discussing these different interpretations of fundamental physics?

Well, in the standard interpretation (which is the predominant view in physics today) time is simply what clocks measure. So when physicists say "time slows down" as we approach the speed of light, they just mean "all clocks in that reference frame slow down".

What about the Neo-Lorentzian view? It agrees with the standard interpretation that as we approach the speed of light, all observable clocks (in that reference frame) slow down. However, that's not what "time" is on this view. On the Neo-Lorentzian view, time is what a clock measures in the undetectable reference frame where the laws of physics work differently than everywhere else in the universe.

Notice how the two theories agree on what happens to observable clocks, the only difference is that the Neo-Lorentzian view simply assumes that there is an undetectable privileged reference frame and that time is what we would measure there, if we could actually measure a clock there.

This causes a few problems for Dr. Craig. First notice how any notion of "time" is completely removed from anything we can observe, but it is still necessarily a very physical entity. This reference frame is a physical place where, on the Neo-Lorentzian view, velocity relative to it has dramatic effects on the material universe that we do observe.

In what way can it be said that we "experience" the flow of time in a physical reference frame that we have no access to? I've covered this already in responding to Craig's "strong argument" for the experience of time, so I won't spend more effort on it here.

The bigger problem comes down to what happens when Craig assumes that this physical form of time has a beginning in the Kalam.

Consider for a moment that Craig is right and that the A-Theory of time is true, and the physical quantity of time described above has a beginning in the finite past. How exactly do we get that from an "eternal" god?

Craig's response here is that god is supposed to be "timeless" before creating time. But this would mean that god is "changeless" before creating physical time, and as such how could we have a period of "eternity" before the universe is created? Craig answers that "god willed from eternity to create the universe", which is a pretty strange answer. On closer inspection if god eternally wills to create time, then how is it possible that the universe is not as old as god? The only way to avoid this problem is to re-introduce the concept of time through the back door: what Craig calls "metaphysical time".

Can you guess what "metaphysical time" is supposed to be? If you went with the Sunday School answer of "Jesus" you're not far off! To Craig, "metaphysical time" is defined by god's sequence of mental events! He even uses the example of god counting down from eternity until "3…2…1…Let there be light!"

This seems to re-introduce the entire problem the Kalam is supposed to solve all over again. Does "metaphysical time" have to have a beginning? Does god have a "first thought?" In fact, if God was counting down from eternity until a finite time ago to create the beginning of "physical time", wouldn't that entail an "actual infinity", which Craig uses philosophical arguments to say can't possibly exist?

Well if we look at Craig's attempted solution to the challenge of an omniscient god knowing an "actual infinity" of things, he tries to say that God's omniscience entails that he knows all true things non-propositionally, as if by instinct. That seems like a very strange way to cash out omniscience, and it has striking similarities to what the B-Theory says about the nature of time, but that's not Craig's biggest problem. The issue is that if God's knowledge is like this, then it is not a mental sequence of events, which is necessary to establish metaphysical time.

What is most surprising about this problem is that this is the exact kind of problem Craig brings up against any kind of a quantum cosmology giving rise to the classical picture of space-time. He insists that if such a quantum cosmology exists before the first Planck second, then it would have produced a universe far sooner than 13.7 billion years ago. Except by the same logic he uses to say *that*, he has the exact same problem with a god "eternally willing to create a universe".

Quite simply, Craig can't have it both ways.

Much like his stances on science and time, he needs a double standard in order for his arguments to work.

PART SIX
Potential Objections

You can see why I thought it was important to devote a good amount of time and space (yes, pun intended) to Counter Apologist, who has had previous encounters with William Lane Craig over these matters. The cherry-picking nature of Craig and his approach to the Kalam is a fundamental flaw that needs to be set out robustly. Now we turn to some objections that might arise.

Although I have already set out some objections within the context of presenting the issues with the Kalam Cosmological Argument here, it would only be fair and rigorous to set out what I think an apologist philosopher such as Craig might claim as counter-arguments. I will then set out to defend my arguments form such objections.

Let me, for ease of reference, lay out my main points of the book so far in a concise manner:

The Form
1) the first premise is inductive and thus the argument can only be as strong as its inductive premise

Premise 1
1) Causality as the universe itself (making the syllogism incoherent or circular)
2) Everything being the universe itself (making the syllogism incoherent or circular)
3) No things begin to exist since things are subjective abstracta which are causally inert

Premise 2
1) the second premise is inductive and thus the argument can only be as strong as its inductive premise

2) *ex nihilo nihilo fit* is a bare assertion with no guarantee it would apply to non-spatiotemporal dimensions
3) Incoherence of even God creating *ex nihilo*
4) Conflation of infinite density with 'nothing'
5) Universe as brute fact is more probable, given Ockham's Razor, than God as a brute fact
6) That the initial singularity on the Standard Big Bang Theory and the BGV are correct in order to imply a definite beginning, and that alternate theories have explanatory power
7) It is more appropriate to remain agnostic over cosmology in this period of theoretic nascence
8) Naturalism is a safer bet then theism for explanatory power and scope in line with prior probabilities

The Conclusion

1) Causality in an a-temporal framework is incoherent
2) Simultaneous causation is not possible
3) Intentionality in an a-temporal framework is incoherent
4) Simultaneous intentionality and action to change a state of affairs is not possible
5) A perfect good would not intend to create and thereby degrade perfection
6) Simultaneity is itself a temporal ideal

In setting out the objections in such a manner, it prompts amazement that so many problems can be associated with three such short lines!

6.1 The Form

In his recent essay on the Kalam Cosmological Argument, set out in the collection of essays which he edited himself with Paul Copan, *Come Let Us Reason: New Essays in Christian Apologetics*, Craig did nothing to refute the accusations made earlier about the form of the argument. Instead, in "Objections So Bad I Couldn't Have Made Them Up", he states in reference to the KCA and the Socrates argument[110]:

> Biological and medical evidence may be marshalled on behalf of the premise that all men are mortal, and I have presented arguments (to be reviewed shortly) for the truth of the premise that everything begins to exist has a cause.

Craig is to some degree denying the definitional quality of the term "man" to not include mortality and refers to medical and biological evidence that men are mortal. However, even if he denies the deductive definitional conclusion that all men are mortal, he is accepting it as inductive, based on evidence. This means that the Socrates argument should be expanded as the KCA was in my first objection:

1) All men that we have observed so far have been mortal
2) Therefore, all men are mortal
3) Socrates is a man
4) Therefore, Socrates is mortal

This argument then becomes a mixture of induction and deduction, and its strength is only as good as the observational evidence to support premise 1. It then follows that either the Socrates argument is definitionally deductive, but the KCA is not, depending on induction for both premise 1 and premise 2, or, as Craig implies, *both* arguments have aspects of induction. Thus, on this most minor of my points, Craig seems to accept, or cannot deny, the inductive quality of the premises.

6.2 Premise 1

Craig does deal with much of the basis to my objections to premise 1 in some of his literature and talks, though does not give it the required rigour or respect. His main defence is in attacking such nominalistic ruminations as deferring to *mereological nihilism*, which he proffers in both his talk at Biola University mentioned earlier and in Craig (2012).

Craig starts[111] by alluding to the fallacy of composition by presenting the objection that just because all the "stuff" within the universe is made of pre-existing matter doesn't imply that the universe itself has always existed. To this I would simply reply that the universe *is* everything. It's not like the universe is a box of marbles and I am applying the same behaviour to the marbles as to the box. This could be problematic. I am saying there is no box, just marbles, or to be more exact, a marble. In this way, there is no behaviour of the marbles, there is just one marble and we cannot infer a model of behaviour from a single event and then apply it to that event. In this way, Craig has not quite got to the heart of the matter.

However, he does claim that this sort of philosophy, in not differentiating matter into discrete units, is a form of mereological nihilism, "the doctrine that there are no composite objects... just fundamental particles which may be arranged chair-wise or table-wise or people-wise or so on"[112]. In such a way, tables and chairs and people do not exist. Craig does not treat this thesis professionally, calling adherents "hapless objectors" and emotively pouring derision upon such thoughts. The term "exist" can itself be a difficult term to accurately define. Does mere mental existence in a single mind denote "existence" in the objective sense that is often implied by the use of the term?

Craig tries to belittle this position (of nominalism over labels and entities) by claiming that this logic means he has always existed. "What was I doing during the Jurassic period?"[113] Of course, it is not that he has *always* existed, it is that William Lane Craig, as an objective thing, has *never* existed. What William Lane Craig *is*, is up for grabs. We can agree, potentially, by consensus, that he is a homo sapiens sapiens, but I have shown (alluding to the Sorites Paradox) that even this is arbitrary and subjective. When he, as a human being, began to exist is impossible to pin down because it is subjective and arbitrary too; this being the entire basis of debate about abortion—when does human personhood begin? At conception? Birth? In between? Production of sperm and egg? And so on. The properties which define William Lane Craig may well *emerge* out of the matter which appears to make up said person, but these emergent properties themselves demand explanation as to their ontology. This becomes an argument over existence properties: What are the essential properties of a chair/person/mountain? Since no one can agree on this complex debate, it is perhaps safe to say that these are subjective and abstract claims about abstract ideas and labels. Do essential properties have any causal power? Can they begin to exist, and if so, when and how? Even if we could establish, without doubt, the beginning to exist, objectively, of a table, this would only be the beginning of an abstract label. To then apply this to the universe itself is a category error, and thus fallacious. Craig does no intellectual legwork in establishing his assertion that mereological nihilism or any other such doctrine is false. All he really offers is the derisory claim "In any case, it is absurd to deny one's own existence" which is poor philosophy when the terms "one" and "existence" are so demanding of further analysis. Are these material or abstract objects? What begins to exist? What supporting evidence and argument does he have for this? And so on.

Without fully understanding human (or any) consciousness, and thus the ontology and delineation of "I", it is difficult to accept jokes about "I" existing or not. There are some incredibly fundamental philosophical questions to be answered in this analysis, and I can see Craig do nothing but sidestep the issue with rather inadequate quips. This is a core question in philosophy that many serious philosophers grapple with, and which still remains open: the question of personal identity and its endurance over time. The me which exists now is, I would claim, different to the me which existed five minutes ago, an hour, a day, a month, a year, twenty years ago and so on. Am I the same person that I was when I was five? When I was a blastocyst? Indeed, some would correctly sense a great similarity to the Greek thought experiment of the Ship of Theseus, whereby every plank and molecule of a ship is replaced, and the question as to whether or not that ship remains the same ship.

So, am I the same person as I was as a blastocyst or a three-year-old?

I would argue not (and in a sense that personhood is a constructed conceptual mechanism which refers to *properties* which *do* exist, in some way, objectively). Could a collection of cells which might end up forming, over time, and with addition of any number of chemicals and processes, be reasonably labelled as Jonathan MS Pearce? With Craig's logic, yes. I would imagine he would be duty-bound to accept that something connects all stages of me with some kind of me-ness. That some kind of Jonathan MS Pearce-ness obtains and maintains over time. This is problematic for a number of reasons, even though it seems commonsensical. This is where the soul seems attractive as a concept because it is this nebulous idea which can supposedly be attached to a bundle of cells, or a grown human, and provide the continuity so desired by theists. Of course, what happens when cells (even up to late blastocyst stage) split and form identical twins? What happens when fertilised eggs are frozen in vitro? And so on. In fact, I could give you a large list of issues with the notion of a soul (as I have written about before) which render the concept rather incoherent. What does a soul *do* or what properties does it *have* that give it any meaningful ontology or function? If it is sentient and can have control over decisions, in what way is it not just consciousness? If a soul is what exists in heaven, does it represent the human at the point of death, when it might have amnesia or senility or dementia, or at an arbitrarily earlier point?

Indeed, the same problem exists with regard to disposition, or evilness if you will. If one is evil (say, a mass murderer) at death, but had been a wonderful, God-fearing Christian all their long lives, then what represents them in soul form (irrespective as to whether they get into heaven or not)? And vice versa, if they were evil all their lives and then repented, sought forgiveness and became Christian in faith and deed at the end of a long wickedness, does the soul represent them in a benign way?

And so on. This is a digression from the central points to be made, but an interesting and relevant one. We can see that things aren't static and objectively definable or objectively existent—a chair, a human being, a hero—and yet we, as thinking conceivers, assume them so, and ascribe to their existences beginnings. It looks, however, to be a wholly conceptual exercise as we attach concrete-seeming labels to arbitrary parts of this ever moving and changing causal and material soup.

To return to the main issue, Craig fails to solve these riddles, instead preferring to lump ridicule on proponents of such positions, or even people who dare question such apparently intuitive notions. In sum, Craig's approach is nothing more than an appeal to ridicule (also known as Stewart's fallacy).

Craig continues by opining:

> At any rate, all these mental machinations are
> ultimately of no avail, for we can simply rephrase the

kalam cosmological argument to meet the scruples of the mereological nihilist:

1`. If the fundamental particles arranged universe-wise began to exist, they have a cause

2`. The fundamental particles arranged universe-wise began to exist.

3`. Therefore, the fundamental particles arranged universe-wise have a cause.

Of all of the things Craig has said about the KCA, this is perhaps the worst. All Craig has done here is assert the following, which I pointed out was one of the main problems with the KCA earlier in this book. He is merely now asserting something about the fundamental arrangement of the particles, and not about them in further arrangements throughout time. So, really, it comes back to this circular reformulation:

1) If the universe began to exist, it has a cause
2) The universe began to exist
3) Therefore, the universe has a cause

Craig is simply making an assertion that the universe began to exist and had a cause, that this fundamental arrangement exists and by doing so is caused. In his original formulation, at least he appealed to the behaviour of something "other than" the universe (i.e. everything) in order to ascribe that behaviour to the universe itself. As we have seen, that fails as an assumption. But here, Craig simply appeals to nothing. It becomes an assumption based on nothing.

If A then B
A
Therefore B

The conditional proposition in premise 1 is an assertion based on nothing; he merely assumes that the apodosis (then it has a cause) follows necessarily from the protasis (if…), which Craig has no observational evidence of. But this is exactly what we are trying to find out! Craig has created a circular argument. He claims that the universe definitely had a start (which we should remain agnostic about, at best), but he cannot appeal to the idea that this start must have had a cause other than a sort of intuition, which is not good enough for an argument that he ascribes such potent and deductive qualities to.

Finally in this section, it is worth noting something else problematic in Craig's thinking, referring to causal principles. When it comes to *his* preferred causal principle, he says, the fact that "everything within the universe which begins to exist has a cause" is evidence that the universe's beginning has a cause. But when it comes to material causation, he says, the fact that everything within the universe is made of pre-existing matter doesn't imply that the universe itself has always existed. But why should we believe one rather than the other? Following Craig, we could defend the proposition, "Anything which begins to exist comes from pre-existing matter," by appealing both to metaphysical intuition and to enumerative induction. Why are we supposed to agree with one metaphysical intuition and enumerative induction and not the other?

In other words, Craig is cherry picking his usage of both intuition and induction.

6.3 Premise 2

After the previous rather unconvincing attempted objection to the problems with the KCA, Craig moves on to differentiating between material and efficient causes. He claims objections such as the ones found here to premise 1 have a "presupposition that everything that begins to exist has a material cause"[114].

Let us first define material and efficient causes. Aristotle talked of four causes: material, formal, efficient and final causes. We should be content with concerning ourselves in light of Craig's defence with two of them. A material cause for an object is the material from which that object is made, such that the cause of a table is the matter and energy from which it is made.[115]

An efficient cause is the primary source which causes the object to come into being. If we were to use the example of a painting, then the material causes would be the canvas and the pigments and paint whilst the efficient cause would be the painter and the art of painting. Craig claims[116]:

> It is true that in our experience material things do not begin to exist without material causes, so that we have the same sort of inductive evidence on behalf of material causation as we have for efficient causation. But if we have good arguments and evidence that the material realm had an absolute beginning preceded by nothing, this can override the inductive evidence. What we cannot reasonably say, I think, is that the universe sprang into being without either an efficient or a material cause, since being does not come from nonbeing. But there is no sort of metaphysical absurdity involved in something's having an efficient cause but no material cause.

What Craig is claiming is that it is clear that, to him, everything in the universe requires material causation until it is inconvenient for him that this is so. If there is no material causation, then efficient causation can step in, if the arguments are good enough. Of course, on inductive evidence and using Craig's understanding of causality, this has never been witnessed. Apparently, God can break metaphysical laws. It would seem to be a more coherent thesis that the universe has existed in some sort of eternal form (cyclical, with time restarting to get around temporal regresses, for example) to mean that the assertion that *ex nihilo nihilo fit* remains intact. This obviously does not suit Craig since he needs to find cosmological support for God as the primary source. We return, then, to Ockham's Razor and the postulation that a brute fact universe is a more simple explanation for the universe than the universe created by a brute fact God (where entities are multiplied unnecessarily). As Hume said, "The beginning of motion in matter itself is as conceivable a priori as its communication from mind and intelligence."[117]

The fact that Craig claims there is "no absurdity" in allowing for efficient causation to replace material causation on its own, despite the entirety of inductive evidence, is nothing more than a convenient assertion. Craig repeats this point elsewhere to a similar objection to the KCA[118]:

> In your final paragraph you appeal to our normal experience of seeing efficient causes acting in tandem with material causes as justification for (1′). But why think that this common concatenation must always be the case?

But this same critique can be applied to his own argument. Why think that this common idea that nothing comes from nothing always be the case? Therefore, Craig seems to fall victim to his own double standards. Craig goes on to give examples of where abstract objects do not need material causes, such as "the equator, the center of mass of the solar system, Beethoven's Fifth Symphony, Leo Tolstoy's *Anna Karenina*, and so forth"[119]. But this is talking about efficient causes for abstract things (ideas and concepts or labels), not efficient causes for material things! It appears to be a rather obvious sidestep. As Hume also states[120]:

> If I am still to remain in utter ignorance of causes, and
> can absolutely give an explanation of nothing, I shall
> never esteem it any advantage to shove off for a
> moment a difficulty which... must immediately, in its
> full force, recur upon me... It were better, therefore,
> never to look beyond the present material world.

Another issue raised within this context is that causality is seen as an interaction between two existing things (material and/or efficient) but it seems incoherent to claim that causality can exist between an existent thing (i.e. God) and a non-existent thing (i.e. the yet-to-be-created universe). In order to interact causally with the universe, therefore, the universe must already exist.

In another simple way, the Kalam can be cut down to size since the term "everything" must necessarily exclude God. Otherwise:

1) Everything requires a cause. To avoid infinite regress, there must be a first cause.

2) But this first cause is something that has no cause.

3) Therefore not everything requires a cause.

4) Therefore the premise is invalid.

Craig would, it seems, appeal to the notion that "everything" excludes God, but it certainly does appear to be a case of special pleading.

6.4 Such views on causality undercut scientific inquiry

One criticism I have encountered whilst putting forth this series of arguments is that this view of causality actually refutes scientific realism. The idea that one cannot cut causality up into chunks for fear of it being arbitrary and subjective means that when we posit something like "plasmodium falciparum causes malaria in human beings" the claim is false. This is because the claim should be "the causal circumstance at C causes malaria in human being X" as well as "the causal circumstance at C` causes malaria in human being X'", and so on. Science (including the social sciences) does seem to rely on arbitrarily breaking off causality into chunks, as I have previously criticised. When the social scientist asks "What causes this particular group of society to be more predisposed to criminality?" there is an implicit and particular understanding (albeit not technically correct) of causality. In both the malaria and criminality cases, we have a tendency to rely on a pragmatic approach to causality such that we *do* cut up our continua of causality. On a simplified model (since causality is a matrix of causality making up the circumstance, with feedback loops and suchlike) we might have:

A causes B, which causes C, which causes D, which causes E

The critic of my approach will say that my claim is that D does not cause E, but science *does* claim this and, as such, I must be a scientific anti-realist. Without wanting to make this part of the discussion too complex, or long, for the purposes of this book, let me say that the critic has a point *to a degree*. If accuracy was to be achieved, one should say that A–D causes E. We do not live in a universe where you can take things *out* of their causal circumstances and test them. We understand language of causal proximity, whereby the nearest tranche of causality to an effect is the cause. These effects are part of one large effect—the universe. If we could extract D and E out of this universe, then we could test for such causality. We would, in effect, be making a small universe out of D and E, but this would prompt the questions as to whether D would actually still cause E in that new, smaller universe.

So when we see that a majority of criminals in such and such a demographic happen to have a correlation of poor education and lack of aspiration, the social scientist might say that poor education and low aspirations *cause* criminality, on balance. Of course, there being exceptions to the rule goes some way to proving my point. It is pragmatically true for the purposes of the study, but in reality, every case of criminality in that study is caused by every single causal circumstance, all of which differ, but all of which are, to that entity (itself arguably arbitrarily cut off from every other entity), unique. One can make a generalised rule from that under the understanding that it is not strictly accurate but that it serves a pragmatic purpose. As the Stanford Encyclopedia of Philosophy states[121]:

> Peirce... holds that the meaning of a proposition is given by its 'practical consequences' for human experience, such as implications for observation or problem-solving. For James... positive utility measured in these terms is the very marker of truth (where truth is whatever will be agreed in the ideal limit of scientific inquiry). Many of the points disputed by realists and antirealists—differences in epistemic commitment to scientific entities, properties, and relations based on observability, for example—are effectively non-issues on this view.

Philosopher J.L. Mackie, in his book *The Cement of the Universe: A Study of Causation*, wrote about INUS conditions with regard to causality, and this is closely connected to my points here (and which I referenced earlier). To clarify as concisely as possible, INUS stands for this sizeable mouthful: *insufficient* but *nonredundant* part of an *unnecessary* but *sufficient* condition. What this means is that one cannot say that a short-circuit caused the house-fire, or that short-circuits cause house-fires; in such a circumstance, the short-circuit was an insufficient cause. On its own, the short-circuit would not have caused the fire, and yet without it, the fire would not have started. This, to me, speaks of all causality within the universe and without things like the Big Bang, and without the whole gamut of preceding "causes", there would be no present effect. When talking about the generalised cause and effect relationships mentioned above, it seems that all the INUS conditions are implicit within any such statement. When someone states, "If you hit the billiard ball at 35 degrees with X Newtons of force towards A, it will do C" implies that the universe exists from the Big Bang or similar, that *you* exist to hit the ball causally and so on. Obviously, to explain the entire causal circumstance every time we make a generalised observation or "rule" about causality is a rather tedious exercise, but this is what is implied. This causal circumstance that Mackie calls a *causal field* hides many background assumptions—it is the entire universe which causes an effect and actually acts as a causal field. Let's look at that short circuit. Think of how many of those INUS conditions had to be in place that led to the fire in the house. The universe had to have formed in such a way that Earth was formed. Humans had to have evolved in such a way that those components were invented and found themselves in that particular house at that particular time, and so on. To say A caused B in a really simplistic manner is useful for shortcuts in pragmatically communicating ideas to each other, but to all other intents and purposes is wrong.

PART SEVEN
Conclusion

The Kalam Cosmological Argument is a long-standing tool employed by a variety of apologists, though William Lane Craig is most prominent in defending it, or more accurately, proposing it. For such a short set of concise lines, there is certainly much that can be said about it. In fact, there is much more that has been defined and explained further here. For example, what is causality, exactly? How does it instantiate itself within objects, or within the relationship between objects? And how does this affect the KCA? Are there different types of causality, and which type is relevant for this syllogism? What of nominalism and realism; which one is fundamentally more coherent? These are further questions which can be taken on from this book, and that many philosophers have indeed endeavoured to answer. I have certainly made some assertions and assumptions of my own without going right to the bottom. But I am telling you this, and laying it out. I *do* think nominalism is more coherent. And so, it seems, does Craig these days. How does his change of view affect his understanding of the KCA? What is he not telling us?

William Lane Craig and his team produce a lot of writing and debates and podcasts, and I don't profess to having read them all. There may be relevant pieces of his work that I have not been aware of and I would happily react to any further defences and points in future dealings with the KCA.

God may have created the universe from nothing. I expect we can never fully prove or disprove such unfalsifiable claims. Indeed, the real aim of this book was not to disprove that God created the universe but to show that the KCA cannot prove that God did, using those premises and the resulting conclusion. I have called into question the terminology used in Craig's formulation of the argument as well as showing that the form is not as deductive as Craig likes to claim.

At the heart of what has been posited is the idea that causality can be seen in discrete units and this is something I hope to have shown does not work. Causality is the causal circumstance of the entire universe since its "creation" event. Thus to use such causality to prove that "the universe" required a cause is unsound. You cannot use an assumed rule of one to prove itself. We have never seen a creation event of a universe, so we assume it requires causality and use that assumption to prove that the universe needs causation!

Time also presents another issue, especially with regard to a creation *ex nihilo* and in partnership with space. Creation of the universe, for God, is more problematic than at first one might think.

These problems exist even if one agrees that the universe did have a finite beginning preceded by nothing, as Craig demands of cosmology. This claim is not met with universal cosmological acquiescence, not by a long shot. There are viable cyclic models and loop quantum cosmology, for example, is enjoying a huge amount of active work and successes, including solving some cosmological problems (resolution of gravitational singularities, a possible mechanism for cosmic inflation and so on). These models *do not* require a finite beginning and thus a First Mover.

Therefore, all told, the Kalam Cosmological Argument does not enjoy the accolades heaped upon it but requires, at the very least, a huge amount of philosophy to be established in order for the premises to even begin to make sense. In the cold light of day, though, it appears rather obvious that the KCA is dead. Demised. Bereft of philosophical life. I am not sure that syllogisms can have miraculous resurrections.

NOTES

---

[1] Smith (2001)

[2] My preference is to capitalise Kalam and to write it without the accent. It is written variously, and quoted thusly, by other writers.

[3] (Craig 2008: 96)

[4] See the Guardian article "Why I refuse to Debate with William Lane Craig", Richard Dawkins, 20/10/2011, http://www.guardian.co.uk/commentisfree/2011/oct/20/richard-dawkins-william-lane-craig (Accessed 13/08/2012)

[5] DeWeese (2011: 29)

[6] Craig (2008) - Part of the (cover) endorsement that Moreland gave for the book.

[7] Craig (2008: 111)

[8] In debates and talks such as his 2010 talk to Biola University entitled "The World's Ten Worst Objections to the Kalam Cosmological Argument", https://www.youtube.com/watch?v=sfsYhWNMYr4 (Accessed 09/12/2015)

[9] Cavender and Kaahane (2009: 37)

[10] Lowder (2013)

[11] Jung (2010)

[12] Bishop (2000: 59-62)

[13] From his 2010 talk to Biola University entitled "The World's Ten Worst Objections to the Kalam Cosmological Argument", linked in [3].

---

[14] I will be using Craig as a synonym for "the author of the KCA".

[15] Martin and Bernard (2002: 95-108)

[16] Morriston gives the reference as quoted in the bibliography here, but the link is no longer valid. I found virtually the same quote in Craig and Moreland (2009: 189)

[17] Craig (2007)

[18] Personally, I find JL Mackie's INUS conditions (insufficient but non-redundant parts of a condition which is itself unnecessary but sufficient for the occurrence of the effect) an interesting concept within the discipline of causality.

[19] One could run down a rabbit hole here in assessing (partial) moral responsibility in light of being part of a causal chain, as can be seen in the work of determinists and compatibilists in the free will debate.

[20] As I have hinted at in the previous note, this has ramifications upon ideas of moral responsibility. As this is not the remit for this book, I would refer the reader to the excellent *Living Without Free Will* by Derk Pereboom.

[21] Adolf Grünbaum's objections that he set out in his 1990 essay "The pseudo-problem of creation in physical cosmology ".

[22] Aristotelian realism proposes that universals, such as redness, exist but are contingent upon the objects which instantiate them (such as a red apple).

[23] Wittgenstein, in his later thought, would have claimed meaning in a word from its use. This of course hints at no objective overarching meaning for groups of things, but meaning derived from each individual usage of language in each context. If anything, this plays into

the point I am making. Things only have meaning to the conceiver, thus don't 'exist' objectively outside the mind of the conceiver, as abstract ideas.

[24] This text is variously available online. I picked it up from: http://www.christianforums.com/t7536666/#post56778897 (Accessed 09/12/2015)

[25] Grünbaum (1989)

[26] Rodriguez-Pereyra (2015)

[27] An argument similar to Bradley's Regress can be applied here, with various solutions offered. See Rodriguez-Pereyra (2015).

[28] Rodriguez-Pereyra (2015)

[29] Craig (2008b)

[30] Craig (2015)

[31] Craig (2012)

[32] Chisholm as cited in Flanagan (2003: ix)

[33] Kane (1996: 4)

[34] Paul Russell, Freedom and Moral Sentiment, 1995, p.14

[35] Pearce (2010)

[36] Pearce (2015)

[37] Loftus (2016)

[38] Vilenkin (2007)

[39] Denigris (2015)

[40] Craig (1993b)

[41] Craig (n.d.)

[42] Craig & Moreland (2009)

[43] By private correspondence.

[44] Lindsay (2013)

[45] It is also worth reading Quentin Smith's paper "Infinity and the Past" (Smith 1987) with regard to these ideas.

[46] Craig and Sinclair, "The Kalam Cosmological Argument", in Craig & Moreland (2009: 103)

[47] William Lane Craig, "Reply to Smith: On the Finitude of the Past", International Philosophical Quarterly 33 (1993).

[48] Craig and Sinclair, "The Kalam Cosmological Argument", in Craig & Moreland (2009: 109)

[49] ibid., 109.

[50] ibid., 111.

[51] Here, $X \subseteq Y$ means that X is a subset of Y (i.e. that all members of X are members of Y), $|X|$ denotes the cardinality of X (i.e. the number of members of X), and X/Y denotes the set difference of X by Y (i.e. all the members of X that are not also members of Y).

[52] Theorem 3 may be proved for an arbitrary countably infinite set $A = \{a_1, a_2, a_3, ...\}$ by taking $B = A$ and $C = \{a_2, a_3, a_4, ...\}$, and noting that $|A| = |B| = |C|$, while $|A/B| = 0 \neq 1 = |A/C|$. The above proof works for

any Dedekind infinite set (a set is Dedekind infinite if it is in one-one correspondence with a proper subset or, equivalently, if it contains a countably infinite subset). Any countably infinite set is Dedekind infinite. The Axiom of Countable Choice implies that any infinite set is Dedekind infinite.

[53] More generally, if μ is any infinite cardinal, then $\mu - \mu$ cannot be defined, although $\mu - v$ can be defined if v is any (finite or infinite) cardinal satisfying $v < \mu$; in this case, $\mu - v = \mu$.

[54] Craig and Sinclair, "The Kalam Cosmological Argument", in Craig & Moreland (2009: 112)

[55] ibid., 112.

[56] Morriston (2002)

[57] Grünbaum (1989)

[58] Craig (n.d.)

[59] This took place at the debate between the two at North Carolina State University, Raleigh, North Carolina, on March 30th, 2011, where Craig picked up on Krauss' use of nothing, claiming he misrepresents "nothing".

[60] Morriston (2002)

[61] Craig (2008c)

[62] Craig (2011)

[63] Craig (2009: 133)

[64] Stenger (2011: 145)

147

[65] The Origin of the Universe and the Arrow of Time – a talk given on October 17th, 2009 to the Perimeter Institute, Waterloo. https://www.youtube.com/watch?v=rEr-t17m2Fo (Accessed 08/12/2012)

[66] Hawking (2010: 128)

[67] In personal correspondence with the author of the website "Debunking William Lane Craig", http://debunkingwlc.wordpress.com/2012/03/ (accessed 20/08/2012)

[68] Sinclair (2011) - James Sinclair wrote on behalf of Craig, in "Current Cosmology and the Beginning of the Universe",

[69] Vilenkin (2007: 177)

[70] Hallquist (2012)

[71] Vilenkin & Tegmark (2011)

[72] Vilenkin (2007)

[73] Guth (n.d.)

[74] Guth (2001: 483)

[75] Theories include ones proposed by Peter Lynds, the Baum–Frampton model, the Steinhardt–Turok ekpyrotic model and the Einstein-Cartan-Sciama-Kibble theory of gravity.

[76] Which itself can be split into the canonical loop quantum gravity as well as the newer covariant quantum gravity.

[77] See the Loop Quantum Gravity section in Weinstein (2015: 3.2)

[78] Sepehri (2015)

[79] Zyga (2015)

[80] Craig (2002)

[81] Stenger (2011: 130)

[82] Eastman (2010)

[83] Edwards (2001: 23)

[84] Edwards (2001: xix)

[85] Carrier and Wanchick (2006)

[86] Craig & Moreland (2009: 194)

[87] Craig (2008: 152)

[88] Grünbaum (1994)

[89] Craig (2007b)

[90] Ibid.

[91] Ibid.

[92] As formulated on the radio interview on the Premier Christian Radio show *Unbelievable?* Recorded with Justin Brierley on 10th September 2011 in the UK. Available online: http://www.premierchristianradio.com/Shows/Saturday/Unbelievable /Episodes/Unbelievable-10-Sep-2011-William-Lane-Craig-Q-A-Tour-preview (Accessed 09/12/2015). A follow-up analysis of Craig's answer by the questioner is included in the following week's show. http://www.premierchristianradio.com/Shows/Saturday/Unbelievable /Episodes/Unbelievable-17-Sep-2011-Can-an-atheist-believe-in-meaning (Accessed 09/12/2015).

[93] An interesting exposition of this argument, which is drawn upon here, has been written up by Dr. John Danaher – Danaher (2011). This draws on an objection to Craig critically discussed by Justin Schieber on an episode of *Reasonable Doubts* (RD92 - http://freethoughtblogs.com/reasonabledoubts/2011/10/14/episode-92-atheists-in-the-pulpit-with-guest-dan-barker/ - Accessed 09/12/2015)

[94] It is also worth referring to John Searle's work (amongst others) on intentionality such as "direction of fit" and "direction of causation", for example in Searle (1983).

[95] In his interviews with Dr. Robert Lawrence Kuhn - http://www.closertotruth.com/contributor/jp-moreland/profile (Accessed 09/12/2015)

[96] Justin Schieber in debate with Michael Horner, "Does the Christian God Exist?", *Real Atheology*, https://www.youtube.com/watch?v=Zf0VL_cMLXQ (Accessed 19/09/2016)

[97] Iron Chariots Wiki (2013)

[98] World English Dictionary entry for "simultaneous", http://dictionary.reference.com/browse/simultaneous (Accessed 29/08/2012)

[99] Wikipedia entry for "A-Series and B-series", https://en.wikipedia.org/wiki/A-series_and_B-series (Accessed 06/12/2015)

[100] ibid.

[101] Markosian (2014)

[102] Craig & Moreland (2009: 183-184)

[103] William Lane Craig in debate with Sean Carroll, "God and Cosmology" 2014 Greer-Heard Forum - https://www.youtube.com/watch?v=07QUPuZg05I (Accessed 07/12/2015)

[104] Craig (2010)

[105] For example, see Counter Apologist's series "Countering the Kalam", of which parts 3 and 4 are particularly relevant. For example, "Countering the Kalam (3) – No Scientific Evidence", *Counter Apologist*, January 10th, 2013, https://counterapologist.blogspot.co.uk/2013/01/countering-kalam-no-scientific-evidence.html (Accessed 10/08/2014)

[106] Sagan (1997: 169)

[107] William Lane Craig, Responses during James Sinclair's talk, "God and Cosmology" 2014 Greer-Heard Forum - Time 48:00 – 49:15 in http://www.youtube.com/watch?v=NaV_wcTIhRc&list=UUxeJuETtMwyydm5D4J2Qsrw

[108] Leininger (2014)

[109] Results, details and met-data and analysis can be found here: http://philpapers.org/surveys/ (Accessed 07/12/2015)

[110] Craig (2012: 57)

[111] Craig (2012: 60)

[112] ibid

[113] Ibid

[114] Craig (2012: 61)

[115] There are issues here with the label 'table' in the context of the Theseus Paradox (the Ship of Theseus). If one replaces every plank and every part of what makes Theseus' ship over time so that after ten years there is no part of that ship which was part of the 'original' ship, is the ship still the same ship?

[116] Craig (2012: 57)

[117] Hume (1854/2006) in Dialogues Concerning Natural Religion, Part VIII

[118] Craig (2007)

[119] Ibid.

[120] Hume (1854/2006) in Dialogues Concerning Natural Religion, Part IV

[121] Chakravartty (2011)

APPENDIX

Jeffery Jay Lowder (2012), from *The Secular Outpost* blog, http://secularoutpost.infidels.org/2012/06/evidential-argument-from-history-of.html (retrieved 29/08/2012)

Informal Statement of the Argument

If there is a single theme unifying the history of science, it is that naturalistic explanations work. The history of science contains numerous examples of naturalistic explanations replacing supernatural ones and no examples of supernatural explanations replacing naturalistic ones. Indeed, naturalistic explanations have been so successful that even most scientific theists concede that supernatural explanations are, in general, implausible, *even on the assumption that theism is true.* Such explanatory success is antecedently more likely on naturalism--which entails that all supernaturalistic explanations are false--than it is on theism. Thus the history of science is some evidence for naturalism and against theism.[1]

Formal Statement of the Argument

Definitions:

physical entity: the kind of entity studied by physicists or chemists. Examples of physical entities include atoms, molecules, gravitational fields, electromagnetic fields, etc.[2]

causally reducible: X is causally reducible to Y just in case X's causal powers are entirely explainable in terms of the causal powers of Y.[2]

ontologically reducible: X is causally reducible to Y just in case X is nothing but a collection of Ys organized in a certain way.[2]

natural entity: an entity which is either a physical entity or an entity that is ontologically or causally reducible to a physical entity.[2]

153

nature: the spatio-temporal universe of natural entities. *Note: there may be additional entities currently unknown to physics but which may be discovered in the future. If and when such entities are discovered, they may be called physical and natural based on their relationship to known physical or natural entities. Thus, this definition of "nature" may only capture nature as currently understood.*[2]

supernatural person: a person that is not part of nature but can affect nature. Examples of supernatural persons include God, angels, Satan, demons, ghosts, etc.[2]

non-natural entity: any entity that is not a natural entity. There are two kinds of non-natural entities: *personal* and *impersonal*. Personal non-natural entities are supernatural persons or agents. Impersonal non-natural entities are abstract objects.

presumption of naturalism: prior to investigation, the probability that the immediate cause of any given natural event is very high.[2]

modest methodological naturalism: scientific explanations may appeal to the supernatural only as a last resort.[2]

naturalistic explanation: a non-supernatural explanation. *Note: a common misunderstanding is the idea that a "naturalistic explanation" means an explanation based in metaphysical naturalism. That is not how "naturalistic explanation" is used here. Rather, a naturalistic explanation simply means any explanation that does not appeal to supernatural agency.*

B: The Relevant Background Information
1. The universe is intelligible.

E: The Evidence to be Explained
1. So many natural phenomena can be explained naturalistically, i.e., without appeal to supernatural agency.
2. The history of science contains numerous examples of naturalistic explanations replacing supernatural ones and no examples of supernatural explanations replacing naturalistic ones.

Rival Explanatory Hypotheses

T: theism: the hypothesis that there exists an omnipotent, omniscient, and morally perfect person (God) who created the universe.

N: metaphysical naturalism: the hypothesis that the universe is a closed system, which means that nothing that is not part of the natural world affects it.

The Argument Formulated

(1) E is known to be true.

(2) Pr(E | B & N) >! Pr(E | B & **T**).

(3) **T** is not much more probable intrinsically than N.

----

(4) Therefore, other evidence held equal, **T** is probably false.

Defense of (2)

N *entails* that any true explanations must be naturalistic ones. Thus, on the assumption that N is true, we have an *extremely* strong reason to expect that successful scientific explanations will be naturalistic ones. In contrast, if **T** is true, then it *could* have been the case that that successful scientific explanations were supernatural explanations. For example, biology could have discovered that all animals are *not* the relatively modified descendants of a common ancestor; neuroscience could have discovered no correlations at all between human minds and brains, etc. If the history of science were like that, then that would have supported **T** over N. But then the success of science in finding naturalistic explanations must be evidence for N over **T**. How strong is this evidence? I agree with Draper: "the more likely it is that there are true naturalistic explanations for natural phenomena (i.e., the stronger the presumption of naturalism), the more unlikely it is that there are supernatural beings."[3]

Objections to AHS

Objections to (2)

Objection: AHS "depends on conflating the old pagan religions and animisms with the Abrahamic religious beliefs. . . . [T]here is a clear distinction between those who believed in the old gods and spirits, and those who held to the Judeo-Christian notion of a transcendent and eternal Creator God. What ended the attribution of supernatural causes to natural processes wasn't the advent of rationalism through the science but the spread of Christianity and it's adherence to a transcendent Creator God who acted uniquely in history to create a universe that acted in accordance with certain laws and principles."[4]

Reply: This reply confuses the distinction between what we might call the "socio-historical explanation" for E with its "metaphysical explanation." The spread of Christianity is an example of the former, while AHS is focused on the latter. Even if that socio-historical explanation for E is correct, it doesn't follow that **T**, much less *Christian* theism, is correct. The objection notes that the universe acts "in accordance with certain laws and principles." That fact is irrelevant to AHS, which explicitly includes the intelligibility of the universe in its background information. *At best,* the fact of the intelligibility of the universe might provide evidence favoring **T** over N. It does not in any way undermine the claim that, *given that the universe acts in accordance with certain laws and principles,* the fact that science has been so successful in providing natural explanations for natural phenomena is evidence favoring N over **T**. To deny this point is to commit the fallacy of understated evidence.[5]

Objection: "For at least fifteen hundred years countless theologians have

156

developed what I call transcendent agent models of divine action in the world which view God as the primary metaphysical agent of all natural events in a way that is completely consistent with the advance of scientific [naturalistic] explanation in its proper sphere. So once again, and for good measure, the advance of science in its proper sphere which it has rightfully claimed from other disciplines, provides no evidence that those other disciplines do not have their proper spheres. And thus it provides no evidence that theism is false or that naturalism is true."[6]

Reply: On the assumption that theism is true, what reason is there to believe that transcendent agent (TA) models of divine action in the world are true? Is there any reason that is independent of the success of non-supernatural explanations?

Let's define A as the hypothesis that "God is the primary metaphysical agent of all natural events in a way that is completely consistent with the advance of scientific explanation in its proper sphere." A is clearly *logically compatible* with E, but the question is whether A undermines premise (2) of AHS. In order to properly evaluate the evidential impact of A, if any, on AHS, I propose that we treat A as an auxiliary hypothesis (to theism). It follows from the theorem of total probability that:

$$\Pr(A \mid T) = \Pr(A \mid T) \times \Pr(E \mid A \& T) + \Pr(\sim A \mid T) \times \Pr(E \mid T \& \sim A)$$

In the context of explanatory arguments, Draper calls that theorem the "weighted average principle" (WAP).[7] As Draper points out, this formula is an average because $\Pr(A \mid T) + \Pr(\sim A \mid T) = 1$. It is not a simple straight average, however, since those two values may not equal $1/2$; that is why it is a *weighted* average.[8] The higher $\Pr(A \mid T)$, the closer $\Pr(E \mid T)$ will be to $\Pr(E \mid A \& T)$; similarly, the higher $\Pr(\sim A \mid T)$, the closer $\Pr(E \mid T)$ will be to $\Pr(E \mid T \& \sim A)$.[9]

WAP shows that, in order to be successful, an objection to an evidential argument must do more than simply *identify* an auxiliary hypothesis which can explain the data. The objection must also provide an antecedent reason for thinking that A is true, i.e., a reason for thinking that A is more probable given the core hypothesis--in this case, theism--than given the negation of the core hypothesis. Without such a reason, the objection reduces to the fact that E is merely *logically compatible* with the core hypothesis (in this case, T), which is no objection at all to an *evidential* argument.

For this reason, then, this objection is, at best, *incomplete*. It successfully identifies a relevant auxiliary hypothesis (A), but does not (yet) provide an antecedent reason for expecting that hypothesis to be true, on the assumption that theism is true.

Objection: Theologians had Biblical reasons for adopting A.

Reply: "Biblical reasons for adopting a TA model" are evidentially relevant if and only if the pattern of probability relations specified by the Weighted Average Principle (WAP) are satisfied. Unless there is an antecedent reason to think that such Biblical reasons are more probable than not, *on the assumption that theism is true,* there is no reason to think A refutes premise (2) of AHS.

Objection: Theologians also had countless philosophical and theological reasons as well. For instance, TA models flow naturally from a commitment to divine simplicity and atemporality, two mainstays of classical theism. So theologians prior to the rise of science had many reasons to endorse A.

Reply: The claim, "TA models flow naturally from a commitment to divine simplicity and atemporality," is just that: a claim, an assertion, in need of support.

158

Objection: Naturalism is unable to explain some 20th-century scientific discoveries, ranging from what happens to when one shoots a photon into space to certain 20th-century scientific discoveries, such as cosmological fine-tuning.

Reply: First, metaphysical naturalism neither logically entails nor makes probable the claim that metaphysical naturalism itself is the explanation for everything studied by the sciences. Rather, metaphysical naturalism entails that all *true* scientific explanations are non-supernatural explanations. According to (2), the fact that so many true scientific explanations are non-supernatural explanations is antecedently much more probable on the assumption that metaphysical naturalism is true than on the assumption that theism is true.

Similarly, the fact that non-supernatural explanations have replaced supernatural explanations, while no supernatural explanations have replaced non-supernatural explanations, is antecedently much more probable on the assumption that metaphysical naturalism is true than on the assumption that theism is true.

Second, I don't rule out that the possibility that scientific discoveries, such as cosmological fine-tuning, could provide evidence for theism and against naturalism. Indeed, I've blogged on *The Secular Outpost* about <u>Draper's argument from moral agency</u> for theism and against naturalism. I've gone so far as to call that argument the *best* argument for theism. None of this, however, undermines the conclusion of AHS, which is that the history of science is *prima facie* evidence against theism. The words "prima facie" are important because they highlight that the argument assesses the evidential impact of one item of evidence *only*. To put the point another way, AHS doesn't claim to examine the *total* available relevant evidence. It's possible that *both* the argument from moral agency *and* AHS are correct. Indeed, it's possible that both are correct, but the former outweighs the latter!

Notes

[1] See Keith M. Parsons, *Science, Confirmation, and the Theistic Hypothesis* (Ph.D. Dissertation, Kingston, Ontario, Canada: Queen's University, 1986), 46; Paul Draper, "Evolution and the Problem of Evil" in *Philosophy of Religion: An Anthology* (3rd ed., ed. Louis Pojman, Wadsworth, 1997), 223-24; and *idem*, "God, Science, and Naturalism" *Oxford Handbook of Philosophy of Religion* (ed. William Wainwright, Oxford: Oxford University Press, 2004), 38-39; and Barbara Forrest, "Methodological Naturalism and Philosophical Naturalism: Clarifying the Connection" *Philo* 3 (2000): 7-29.

[2] Draper 2004.

[3] Draper 2004.

[4] Jack Hudson, "Arguments for God: The Historically Unique Nature of God." *Wide as the Waters* (August 10,2010), http://jackhudson.wordpress.com/2010/08/10/the-historically-unique-nature-of-creation/ (spotted June 16, 2012).

[5] Paul Draper, "Partisanship and Inquiry in the Philosophy of Religion," unpublished paper. Cf. Draper 2004, 43-44.

[6] Randal Rauser, "A Critical Look at Jeff Lowder's Evidential Argument from the History of Science." *Randal Rauser* (July 13, 2012), http://randalrauser.com/2012/07/a-critical-look-at-jeff-lowders-evidential-argument-from-the-history-of-science/ (spotted July 13, 2012).

[7] Paul Draper, "Pain and Pleasure: An Evidential Problem for Theists" *Noûs* 23 (1989): 331-50 at 340.

[8] Draper 1989.

[9] Draper 1989.

Related Resources

Tyson, Neil deGrasse. "The Perimeter of Ignorance." *Natural History Magazine* (November 2005).

BIBLIOGRAPHY

Bishop, Paul (2000), *Synchronicity and Intellectual Intuition in Kant, Swedenborg, and Jung*, New York: The Edwin Mellen Press

Bokulich, Peter and Curiel, Erik (2009), "Singularities and Black Holes", *The Stanford Encyclopedia of Philosophy*, http://plato.stanford.edu/entries/spacetime-singularities/ (Accessed 09/12/2015)

Carrier, Richard, and Wanchick, Tom (2006), "Naturalism vs. Thesim", http://www.infidels.org/library/modern/richard_carrier/carrier-wanchick/index.html (Accessed 09/12/2015)

Cavendar, Nancy and Kahane, Howard (2009), *Logic and Contemporary Rhetoric: The Use of Reason in Everyday Life*, Belmont: Wadsworth, Cengage Learning

Chakravartty, Anjan (2011), "Scientific Realism", *The Stanford Encyclopedia of Philosophy*, http://plato.stanford.edu/entries/scientific-realism/ (Accessed 09/12/2015)

Craig, William Lane (1979), *The Kalam Cosmological Argument*, London: McMillan Press

Craig, William Lane (1993), "Reply to Smith: On the finitude of the past", *International Philosophical Quarterly*, 33.

Craig, William Lane (1993b), "The Caused Beginning of the Universe: a Response to Quentin Smith." *British Journal for the Philosophy of Science* 44: 623-639.

Craig, William Lane (2001), *God, Time, and Eternity: The Coherence of Theism II: Eternity*, AH Dordecht: Kluwer Academic Publishers

Craig, William Lane (2002), "The Existence of God and the Beginning of the Universe", *Truth Journal*, v. 3, (http://www.iclnet.org/clm/truth/3truth11.html) (Accessed 09/12/2015)

163

Craig, William Lane (2007), "Causal Premiss of the Kalam Argument", http://www.reasonablefaith.org/causal-premiss-of-the-kalam-argument (Accessed 20/08/2012)

Craig, William Lane (2007b), "God, Time, and Creation", http://www.reasonablefaith.org/god-time-and-creation (Accessed 20/08/2012)

Craig, William Lane (3rd ed. 2008), *A Reasonable Faith*, Wheaton, Illinois: Crossway Books

Craig, William Lane (2008b), "Current Work on God and Abstract Objects", http://www.reasonablefaith.org/current-work-on-god-and-abstract-objects (Accessed 09/12/2015)

Craig, William Lane (2008c), "Contemporary Cosmology and the Beginning of the Universe", http://www.reasonablefaith.org/contemporary-cosmology-and-the-beginning-of-the-universe (Accessed 20/08/2012)

Craig, William Lane (2011), "Lawrence Krauss' Response and Perspective", http://www.reasonablefaith.org/lawrence-krauss-response-and-perspective (Accessed 20/08/2012)

Craig, William Lane (2010), "Beginning to Exist", http://www.reasonablefaith.org/beginning-to-exist (Accessed 07/12/2015)

Craig, William Lane (2011b), "Current Cosmology and the Beginning of the Universe", http://www.reasonablefaith.org/current-cosmology-and-the-beginning-of-the-universe (Accessed 20/08/2012)

Craig, William Lane (2012), "Nominalism and Divine Aseity", *Oxford Studies in Philosophy of Religion Volume 4,* (Ed. Kvanvig, Jonathan), Oxford: Oxford University Press, http://www.oxfordscholarship.com/view/10.1093/acprof:oso/9780199656417.001.0001/acprof-9780199656417-chapter-3 (Accessed 09/12/2015)

Craig, William Lane (2015), "Nominalism and Natural Law", http://www.reasonablefaith.org/nominalism-and-natural-law (Accessed 09/12/2015)

Craig, William Lane (n.d.), "The Existence of God and the Beginning of the Universe", http://www.reasonablefaith.org/the-existence-of-god-and-the-beginning-of-the-universe (Accessed 09/12/2015)

Craig, William Lane (n.d.b), "Divine Timelessness and Personhood", http://www.reasonablefaith.org/divine-timelessness-and-personhood

Craig, William Lane and Moreland, J. P. (2009), *The Blackwell Companion to Natural Theology*, Chichester: Blackwell Publishing Ltd.

Craig, William Lane and Copan, Paul (2012), *Come Let Us Reason: New Essays in Chrsitian Apologetics*, Nashville: B & H Publishing Group

Danaher, John (2011), "Schieber's Objection to the Kalam Argument", *Philosophical Disquisitions*, http://philosophicaldisquisitions.blogspot.co.uk/2011/12/schiebers-objection-to-kalam-argument.html (retrieved 28/08/2012)

Denigris, Matthew (2015), "On the Kalam Cosmological Argument", *The Winnower* 3:e144157.70670. DOI: 10.15200/winn.144157.70670

DeWeese, Garrett J. (2011), *Doing Philosophy as a Christian*, Downers Grove, IL: InterVarsity Press

Dowe, Phil (2007), "Causal Processes", *The Stanford Encyclopedia of Philosophy*, http://plato.stanford.edu/entries/causation-process/ (Accessed 09/12/2015)

East, James (2013), "Infinity Minus Infinity", *Faith and Philosophy*, Volume 30, Issue 4, Pages 429-433

Eastman, Timothy E. (2010), "Cosmic Agnosticism, Revisited", *Journal of Cosmology*, 2010, Vol 4, pages 655-663.

Edwards, Rem B. (2001), *What Caused the Big Bang?*, Amsterdam: Editions Rodopi B.V.

Flanagan, Owen (2003), *The Problem Of The Soul: Two Visions of Mind and How to Reconcile Them*, New York: Basic Books

Grünbaum, A. (1990) "Pseudo Creation of the Big Bang" *Nature*, vol 344

Grünbaum, A. (1989) "The Pseudo-Problem of Creation in Physical Cosmology", *Philosophy of Science*, Vol. 56, No. 3, Sept. 1989, pp. 373-394
http://www.infidels.org/library/modern/adolf_grunbaum/problem.html

Grünbaum, A. (1991) "Creation As a Pseudo-Explanation in Current Physical Cosmology", *Erkenntnis* 35: 233-254, 1991.

Grünbaum, A. (1994) "Some Comments on William Craig's "Creation and Big Bang Cosmology", *Philosophia Naturalis*, Vol. 31, No. 2, pp. 225-236, 1994

Guth, Alan (2001), "The Ultimate Free Lunch", *The book of the cosmos*, (ed. Danielson, Dennis Richard), Cambridge, MA: Perseus Books Group

Guth, Alan (n.d.), "Did the Universe Have a Beginning?", http://www.counterbalance.org/cq-guth/didth-frame.html (Accessed 26/11/2015)

Hallquist, Chris (2012), "Victor Stenger on William Lane Craig", http://www.patheos.com/blogs/hallq/2012/08/victor-stenger-on-william-lane-craig/ (Accessed 26/11/15)

Hawking, Stephen (2010), *The Grand Design*, New York: Bantam Books

Hume, David (1854/2006), *Dialogues Concerning Natural Religion*, corrected version of the 1854 original using the Kemp Smith edition, by David Banach:
http://www.anselm.edu/homepage/dbanach/dnr.htm (Accessed 07/12/2015)

Humphreys, Paul (1986), "Causation in the social sciences: An overview", *Synthese*, Volume 68, Number 1, 1-12

Iron Chariots Wiki (2013), "Problem of non-God objects", *Iron Chariots Wiki*, http://wiki.ironchariots.org/index.php?title=Problem_of_non-God_objects (Accessed 13/07/2013)

Jung, Carl (2010), *Synchronicity: An Acausal Connecting Principle. (From Vol. 8. of the Collected Works of C. G. Jung)*, (trans. R.F.C. Hull), Princeton: Princeton University Press

Kane, Robert (1996), *The Significance of Free Will*, New York: Oxford University Press

Leininger, Lisa (2014), "On Mellor and the Future Direction of Time", *Analysis*, 74 (1): 148-157, http://people.ds.cam.ac.uk/dhm11/images/MMRAnalRev.pdf (Accessed 07/12/2015)

Lindsay, James A. (2013), *Dot, Dot, Dot: Infinity Plus God Equals Folly*, Fareham: Onus Books

Loftus, John (2016), *Christianity in the Light of Science: Critically Examining the World's Largest Religion*, Amherst: Prometheus

Lowder, Jeffrey Jay (2013), "How the Distinction between Deductive vs. Inductive Arguments Can Mask Uncertainty", *The Secular Outpost*, http://www.patheos.com/blogs/secularoutpost/2013/01/28/how-the-distinction-between-deductive-vs-inductive-arguments-can-mask-uncertainty/ (Accessed 16/08/2016)

Ma, Yongge (2011), "The Cyclic Universe Driven by Loop Quantum Cosmology", *Journal of Cosmology*, vol 13, 2011 http://journalofcosmology.com/MacyclicUniverse4.pdf

Markosian, Ned (2014, substantive revision from 2002), "Time", *The Stanford Encyclopedia of Philosophy*, http://plato.stanford.edu/entries/time/ (Accessed 06/12/2015)

Mellor, D. H. (1998), *Real Time II*, London, Reoutledge

Morriston, Wes (2002), "A Critique of the Kalam Cosmological Argument" in Martin, Ray and Bernard, Christopher, *God Matters*, New York: Longman

Morriston, Wes (n.d.), "Creation Ex Nihilo and the Big Bang", *Philo*, vol 5 no. 1

Pearce, Jonathan MS (2010), *Free Will? An investigation into whether we have free will or whether I was always going to write this book*, Fareham: GPP

Pearce, Jonathan MS (2015), "Is Society Accepting That Free Will Is an Illusion?", *Free Inquiry*, vol 35, issue 4, May 19th 2015

Pereboom, Derk (2003), *Living Without Free Will*, Cambridge: Cambridge University Press (Virtual Publishing)

Rodriguez-Pereyra, Gonzalo (2015, substantive revision of 2008), "Nominalism in Metaphysics", *The Stanford Encyclopedia of Philosophy*, http://plato.stanford.edu/entries/nominalism-metaphysics/ (Accessed 07/12/2015)

Sagan, Carl (1996), *The Demon-Haunted World*, New York: Ballantine Books

Schaffer, Jonathan (2007), "The Metaphysics of Causation", *The Stanford Encyclopedia of Philosophy*, http://plato.stanford.edu/entries/causation-metaphysics/ (Accessed 07/12/2015)

Searle, John (1983), *Intentionality: An Essay in the Philosophy of Mind*, Cambridge: Cambridge University Press

Sepehri, Alireza (2015), "Cosmology from quantum potential in brane–anti-brane system", *Physics Letters B*, Volume 748, 2 September 2015, Pages 328–335

Sinclair, James (2011), "Current Cosmology and the Beginning of the Universe", on behalf of William Lane Craig, http://www.reasonablefaith.org/current-cosmology-and-the-beginning-of-the-universe (Accessed 20/08/2012)

Smith, Quentin (1987), "Infinity and the Past", *Philosophy of Science*, Vol. 54, No. 1 (Mar., 1987), pp. 63-75

Smith, Quentin (2001),"The Metaphilosophy of Naturalism", *Philo* 4: 195-215.

Still, James, "Eternity and Time in William Lane Craig's Kalam Cosmological Argument", http://www.infidels.org/library/modern/james_still/kala m.html (Accessed 07/12/2015)

Stenger, Victor J. (2011), *The Fallacy of Fine-Tuning: Why the Universe is not Designed for Us*, Amherst, NY: Prometheus

Vilenkin, Alexander (2007), *Many Worlds in One: The Search for Other Universes*, New York: Hill & Wang (Farrar, Straus and Giroux)

Vilenkin, Alexander & Tegmark, Max (2011), "The Case for Parallel Universes: Why the multiverse, crazy as it sounds, is a solid scientific idea", *Scientific American*, http://www.scientificamerican.com/article/multiverse-the-case-for-parallel-universe/ (Accessed 01/12/2015)

Weinstein, Steven (2015, substantive revision from 2005), "Qauntum Gravity", *The Stanford Encyclopedia of Philosophy*, http://plato.stanford.edu/entries/quantum-gravity/ (Accessed 26/11/2015)

Zyga, Lisa (2015), "No Big Bang? Quantum equation predicts universe has no beginning", *Phys.org*, http://phys.org/news/2015-02-big-quantum-equation-universe.html#jCp (Accessed 26/11/2015)

Other books by Jonathan MS Pearce:

Free Will? An investigation into whether we have free will or whether I was always going to write this book

The Nativity: A Critical Examination

The Little Book of Unholy Questions

13 Reasons to Doubt (Ed.)

Beyond an Absence of Faith (Ed.)

The Problem with "God": Classical Theism under the Spotlight

As Johnny Pearce:

Survival of the Fittest: Metamorphosis

CPSIA information can be obtained
at www.ICGtesting.com
Printed in the USA
BVOW08s0824120118
504930BV00001B/73/P